Quantum Physics Made Easy for Beginners

Dante .B Gonzales

<u>*Funny helpful tips:*</u>

Practice patience; good things often take time.

Stay true to the bond; it's a treasure.

Quantum Physics Made Easy for Beginners : Unlock the Secrets of Quantum Physics: A Beginner's Guide to Understanding the Fundamentals

Life advices:

Stay connected with the world of exoskeletons; they promise enhanced mobility and strength augmentation.

Life's tapestry is rich with colors; weave it with diverse experiences and perspectives.

Introduction

This book provides an accessible introduction to the fundamental concepts and applications of quantum mechanics.

The book delves into the intriguing duality of waves and particles, shedding light on the wave-particle duality theory and its implications in the realm of quantum physics. It explores how the power of quantum mechanics revolutionizes our understanding of the microscopic world, enabling the development of groundbreaking technologies and scientific innovations.

Furthermore, the guide delves into the intricate properties of metals and insulators, elucidating their behavior in the context of quantum physics. It discusses the role of semiconductors in the creation of computer chips and their significance in modern technology, emphasizing their crucial role in the advancement of electronic devices.

Moreover, the book provides an in-depth exploration of the concept of superconductivity, unraveling the mysteries behind the phenomenon and its practical applications in various fields. It explains the principles of spin doctoring, offering insights into the manipulation of spin states and their potential impact on quantum computing and information processing.

With its comprehensive coverage of key topics in quantum physics, the book serves as an invaluable resource for beginners seeking to grasp the foundational principles and real-world implications of this complex and fascinating scientific discipline.

Contents

CHAPTER 1: INTRODUCTION

Quantum Physics VS. Rocket Science

In modern years, rocket science has become a byword for something genuinely challenging. Rocket specialists need a thorough understanding of the properties of the materials used in spacecraft construction; they need to understand the ability and risk of the fuels used to power the rockets, and they need a thorough understanding of how planets and satellites are moving under the influence of gravity.

Quantum physics has a similar reputation for complexity, and, even for many highly educated physicists, a thorough understanding of the behaviour of many quantum phenomena definitely poses a significant challenge. Perhaps the best minds in physics are those working on the unsolved issue of how quantum physics can be applied to the incredibly strong gravitational forces that are supposed to exist inside black holes, which played a crucial role in our universe's early evolution.

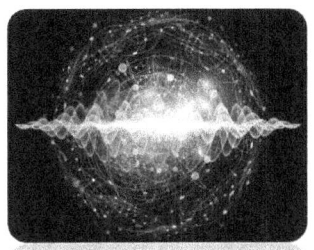

The basic ideas of quantum physics, however, are not rocket science: their problem is more to do with their unfamiliarity than with their inherent difficulty. We have to abandon some of the ideas we all learned from our observation and knowledge of how the world functions, but once we have done so, it is more an exercise for the imagination than the intellect to replace them with the new concepts needed to understand quantum physics.

It is also very easy to understand how many everyday phenomena underlie the concepts of quantum physics without using the complex mathematical research required for full clinical care.

Chapters Overview

The philosophical foundation of quantum physics is peculiar and unfamiliar, and it is still controversial in its interpretation. We will, however, postpone much of our discussion of this to the last chapter since the main purpose of this book is to understand how quantum physics explain many natural phenomena; these include the behavior of matter on the very small scale of atoms and the like, but also many of the phenomena we in the modern world are familiar with.

We shall establish the basic concepts of quantum physics in Chapter 2, where we will find that the fundamental particles of matter are not like ordinary objects, such as footballs or grains of sand, but can, in certain cases, behave as if they were waves. We will find that in deciding the structure and properties of atoms and the 'subatomic' environment beyond them, this 'wave-particle duality' plays an important role.

Chapter 3 starts our discussion of how important and common aspects of everyday life underlie the concepts of quantum physics. This chapter describes how quantum physics is central to many of the techniques used to produce power for modern society, called 'Power from the Quantum.' We can also find that the 'greenhouse effect' is essentially quantum, which plays an important role in regulating the temperature and, thus, our world's climate. Much of our industrial technology contributes to the greenhouse effect, contributing to global warming issues, but quantum physics also plays a role in combating the physics of some of the 'green' technologies being developed.

In Chapter 4, we can see how in some large-scale phenomena, wave-particle duality features; for instance; quantum physics explains why some materials are metals that can conduct electricity, while others are 'insulators' that fully block such current flow.

The physics of 'semi-conductors' whose properties lie between metals and insulators are discussed in Chapter 5. In these materials, which were used to build the silicon chip, we will find out how quantum physics plays an important role. This system forms the basis of modern electronics, which, in turn, underlies the technology of information and communication, which plays such a huge role in the modern world.

We shall turn to the 'superconductivity' phenomenon in Chapter 6, where quantum properties are manifested in a particularly dramatic way: in this case, the large-scale existence of the quantum phenomena creates materials whose resistance to electric current flow disappears entirely. Another intrinsically quantum phenomenon relates to newly established information processing techniques, and some of these will be discussed in Chapter 7.

There, we can discover that it is possible to use quantum physics to relay information in a way that no unauthorized individual can interpret. We can also learn how to construct 'quantum computers' one day to perform certain calculations several millions of times faster than any current machine would.

Chapter 8 tries to bring everything together and make some guesses about where the topic might be going. Most of this book, as we see, relates to the influence of quantum physics on our daily world: by this, we mean phenomena where the quantum component is seen at the level of the phenomenon we are addressing and not just concealed in the quantum substructure of objects. For instance, while quantum physics is important to understand the internal structure of atoms, the atoms themselves follow the same physical laws in many circumstances as those governing the behavior of ordinary objects.

Thus, the atoms move around in gas and clash with the container walls and with each other as if they were very tiny balls. On the other hand, their internal structure is determined by quantum laws when a few atoms come together to form molecules, and these directly control essential properties such as their ability to absorb and re-emit greenhouse effect radiation (Chapter 3).

The context needed to understand the ideas I will build in later chapters is set out in the current chapter. I begin by defining some basic ideas that were established before the quantum era in mathematics and physics; I then offer an account of some of the discoveries of the nineteenth century, especially about the nature of atoms, that revealed the need for a revolution in our thought that became known as 'quantum physics.'

Mathematics

Mathematics poses a major hurdle to their comprehension of science for many individuals. Certainly, for four hundred years or more, mathematics has been the language of physics, and without it, it is impossible to make progress in understanding the physical universe. Why will this be the case? The physical universe seems to be primarily governed by the laws of cause and effect, for one explanation (although these break down to some extent in the quantum context, as we shall see). Mathematics is widely used to evaluate such causal relationships: the mathematical statement two plus two equals four 'implies as a very simple example that if we take any two physical objects and combine them with any two others, we will end up with four objects.

If an apple falls from a tree, to be a little more sophisticated, it will fall to the ground, and we can use mathematics to measure the time it will take, given we know the initial height of the apple and the strength of the gravity force acting on it. This shows the relevance of mathematics to science since the latter attempts to predict and compare the behavior of a physical system with the outcomes of 4 Quantum Physics: measurement.

Classical Physics

If quantum physics is not rocket science, we can also assume that quantum physics is not 'rocket science.' This is because it is possible to measure the motion of the sun and the planets as well as that of rockets and artificial satellites with total precision using pre-quantum physics developed by Newton and others between two and three hundred years ago.

The need for quantum physics was not understood until the end of the nineteenth century because in many familiar situation's quantum effects are far too small to be important. We refer to this earlier body of information as 'classical' when we address quantum physics.

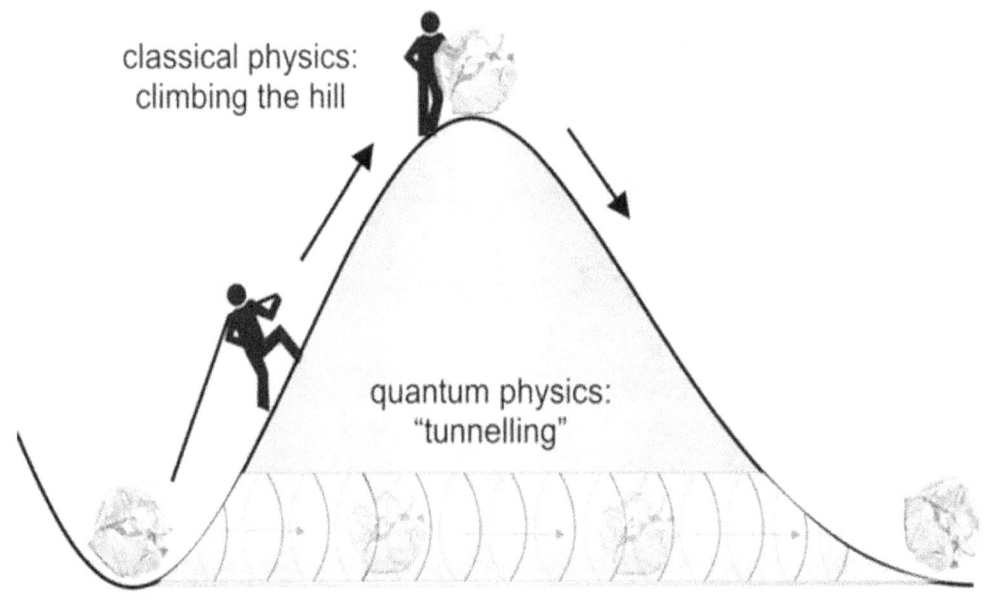

classical physics:
climbing the hill

quantum physics:
"tunnelling"

In some scientific fields, the term 'classical' is used to mean anything like 'what was understood before the subject we are addressing became important,' so it refers to the body of scientific information that preceded the quantum revolution in our sense. The early quantum physicists were acquainted with the notions of classical physics and used them to generate new ideas where they could. We will follow in their footsteps and will soon answer the key ideas of classical physics that will be needed in our subsequent debate.

Units

We have to use a scheme of 'units' when physical quantities are represented by numbers. For instance, we could calculate the distance in miles, in which case the mile would be the unit of distance, and time in hours, where the hour would be the unit of time, and so on. By the French name 'Systeme Internationale' or 'SI' for short, the system of units used in all scientific work is known. The distance unit is the meter (abbreviation 'm') in this system, the time unit is the second ('s'), mass is calculated in kilogram units ('kg'), and the electrical charge is measured in coulomb units ('C').

Unit	Symbol	Quantity
kilogram	kg	mass
metre	m	length
second	s	time
ampere	A	electric current
kelvin	K	temperature
mole	mol	amount of substance
candela	cd	luminous intensity

When the metric system was developed in the late eighteenth and early nineteenth centuries, the dimensions of the fundamental units of mass, length, and time were originally specified. The meter was originally specified as one ten-millionth of the distance from the pole to the equator along the meridian that passes through Paris; the second as 1/86,400 of the average solar day; and the kilogram as one-thousandth of the mass of pure water per cubic meter. These concepts gave rise to problems because our ability to more precisely calculate the dimensions of the Earth and motion meant minor improvements in these standard values.

The meter and kilogram were redefined towards the end of the nineteenth century as, respectively, the distance between two marks on a standard platinum alloy rod and the mass of another particular piece of platinum; both of these standards were kept firmly in a standard laboratory near Paris, and 'secondary standards' were manufactured to be as identical as possible to the originals. In 1960, the definition of the second was updated and expressed in terms of the year's average duration.

As atomic measurements became more precise, the basic units were again redefined: the second is now known as 9,192,631,770 radiation oscillation cycles emitted during the change between the specific energy levels of the cesium atom, while the meter is defined as the distance traveled by light in a time equal to 1/299,792,458 of a second. The value of these concepts is that, everywhere on Earth, the standards can be replicated independently. However, no similar definition of a kilogram has yet been accepted, and this is still referred to as the primary standard kept by the Bureau of Standards of France.

In our labs, kitchens, and elsewhere, the values of the standard masses we use were all obtained by comparing their weights with standard weights, which were compared with others in turn, and so on until we finally reached the Paris standard. The standard unit of charge is measured by means of the ampere, which is the current standard unit and is equal to one coulomb per second. The ampere itself is defined as the current needed between two parallel wires kept one meter apart to generate a magnetic force of a specific size. Other physical quantities are determined in units derived from these four: thus, by dividing the distance traveled by the time taken, the speed of a moving object is estimated, so the unit speed corresponds to one meter divided by one second, which is written as 'ms-1'.

Motion

A large part of physics concerns objects in motion, both classical and quantum, and the simplest definition used here is that of speed. For an object traveling at a steady speed, this is the distance it moves in one second (measured in meters). If the speed of an object changes, then its value is defined at any given time as the distance it would have traveled in one second had its speed remained constant.

For someone who has driven in a motorcar, this concept should be familiar, although the units are usually kilometers (or miles) per hour in this case. That of 'velocity' is closely linked to the idea of speed. Both words are interchangeable in everyday speech, but in physics, they are differentiated by the fact that velocity is a quantity of 'vector,' which means it has both direction and magnitude.

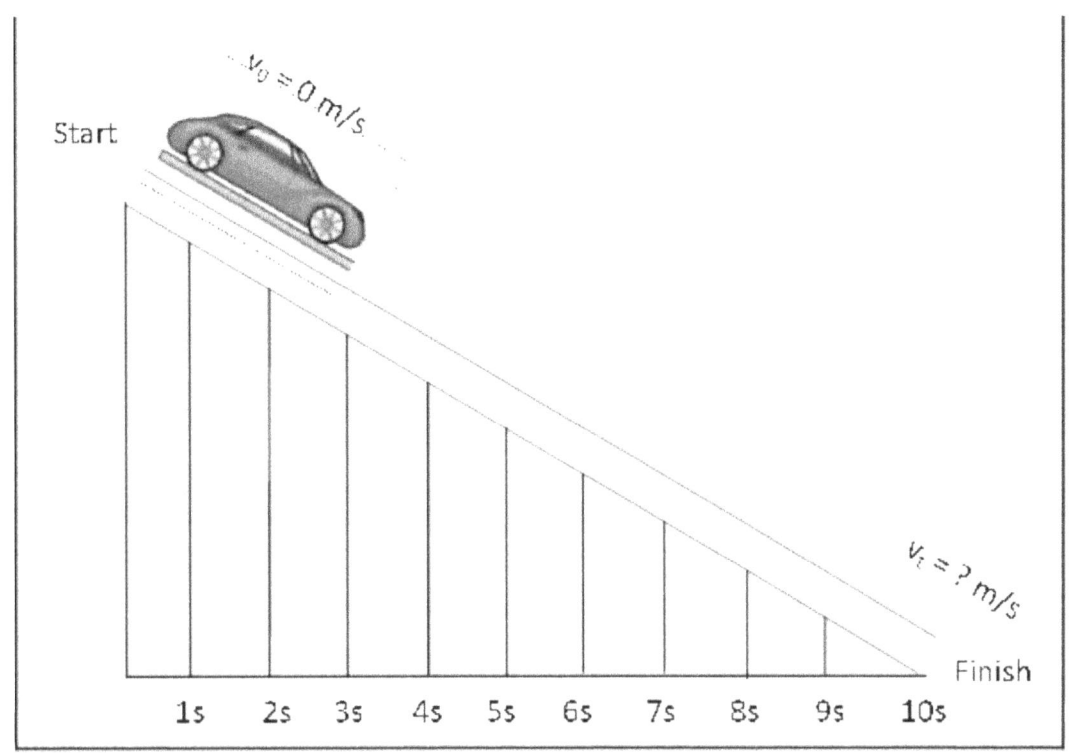

Therefore, an object traveling from left to right at a speed of 5 ms-1 has a five ms-1 positive velocity, but one moving from right to left at the same speed has a five ms-1 negative velocity. The rate at which it does so is known as acceleration, when the velocity of an object is changing. For example, if the speed of an object varies from 10 ms-1 to 11 ms-1 over a span of one second, the velocity shift is 1 ms-1, so its acceleration is '1 meter per second squared' or 1 ms-2.

Mass

The mass of a body was defined by Isaac Newton as 'the amount of matter' it contains, which raises the question of what matter is or how its 'quantity' can be calculated. The problem is that while certain quantities can be described in terms of more simple quantities (e.g., speed in terms of distance and time), some definitions are so important that any such attempt leads to a circular description such as that just stated.

To escape from this, we should 'operationally' identify certain quantities, implying that we explain what they do rather than what they are, i.e., how they function. In the case of mass, when subjected to gravity, this can be achieved by force encountered by an object.

Thus, when positioned at the same point on Earth's surface, two bodies with the same mass can feel the same force, and the masses of two bodies can be measured using a balance.

Energy

In our later discussions, this is an idea we would always refer to. An example is energy possessed by a moving body, defined as 'kinetic energy'; this is measured by the square of its velocity as one-half of the body's mass-so its units are joules, equal to kgm2s-2.

Potential energy, which is related to the force acting on the body, is another essential source of energy. An example is gravity-related potential energy, which increases in proportion to the distance that an object is lifted from the floor. By multiplying the mass of the object by its height and then by the acceleration due to gravity, its weight is determined.

The units of these three quantities are kg, m, and ms-2, respectively, so the potential energy unit is kgm2s2, which is the same as the kinetic energy unit, which is to be expected since it is possible to transfer various sources of energy from one to another.

In both quantum and classical physics, an extremely significant concept is that of 'energy conservation,' which means that it is never possible to produce or destroy energy. It is possible to transform energy from one form to another, but the total quantity of energy is still the same. By considering one of the simplest examples of a physical operation, we can demonstrate this,

An object falls under gravity. If we take some object and drop it, we find that it travels faster and faster when it drops to the ground. As it moves, it decreases its potential energy, increasing its speed and thus its kinetic energy. The total energy is the same at any point.

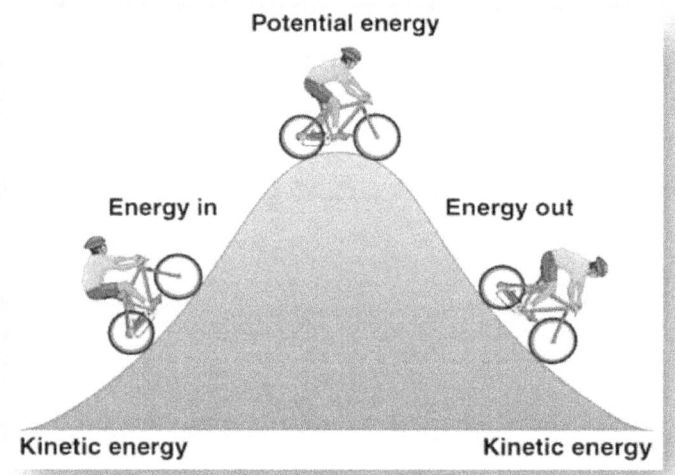

Now imagine what occurs on Earth after the dropping object falls. Assuming it doesn't bounce, both its kinetic and potential energies have diminished to zero, so where has the energy gone?

The reason is that it was turned into heat that warmed up the World around it.

In the case of ordinary objects, this is just a small impact, but the release of energy can be immense when large bodies fall: for instance, the collision of a meteorite with the Earth several million years ago is thought to have contributed to the extinction of dinosaurs. Electrical energy (to which we shall return shortly), chemical energy, and mass-energy are other examples of types of energy as expressed in Einstein's famous equation, $E = mc^2$.

Electric Charge

In classical physics, there are two major sources of potential energy. One is gravity, which we alluded to above, and the other is energy, also called 'electromagnetism' and synonymous with magnetism. Electricity is a fundamental concept of electricity, and, as a mass, it is a quantity that is not readily described in terms of other more fundamental concepts, so we use an operational description again. A force is exerted on each other by two bodies bearing electric charges.

If the charges have the same signal, this force is repulsive and drives the bodies away from each other, while it is enticing and draws them together if the signals are opposite.

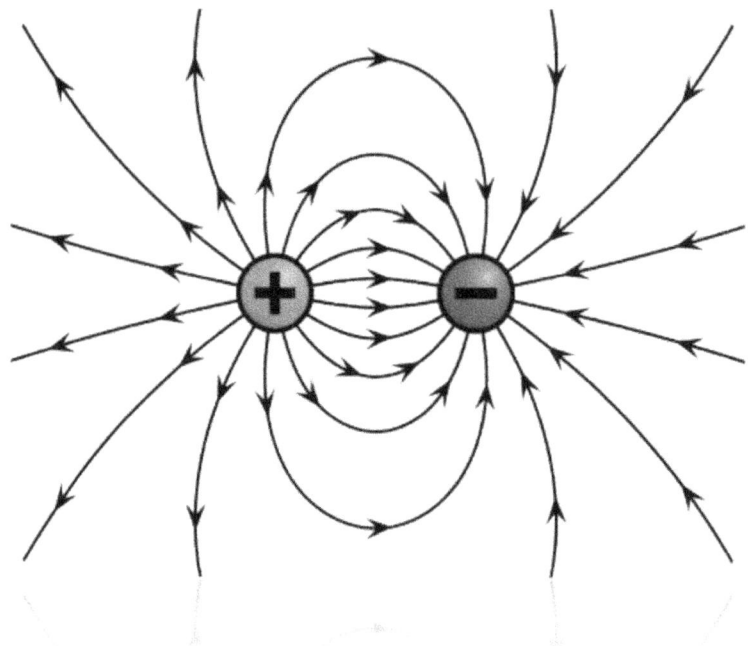

In both situations, they would gain kinetic energy if the bodies were released, flying apart in the like-charge case or together if the charges are opposite. There must be potential energy associated with the interaction between the charges to ensure the energy is conserved, one that gets larger as the related charges come together or as the different charges split.

Not only does the electric field shift as charges pass, but another field, the 'magnetic field,' is formed. Familiar examples of this field are those formed by a magnet or, indeed, by the Earth, which controls the direction of a compass's needle. In the form of 'electromagnetic waves,' one example of which is light waves, the coupled electrical and magnetic fields generated by moving charges propagate through space. In Chapter 2, we shall return to this in more detail.

A moving body's momentum is defined as the product of its mass and its velocity, so a slow-moving heavy object may have the same momentum as a fast-moving light body. The cumulative momentum of both remains the same when two bodies collide, so the momentum is 'preserved' just as in the case of previously mentioned energy. In an important respect, however, momentum is different from energy: it is a vector quantity (like velocity) with both direction and magnitude.

When we drop the ball on the ground, and it bounces upward at around the same speed, the sign of its momentum changes such that the cumulative change in momentum equals its initial value twice.

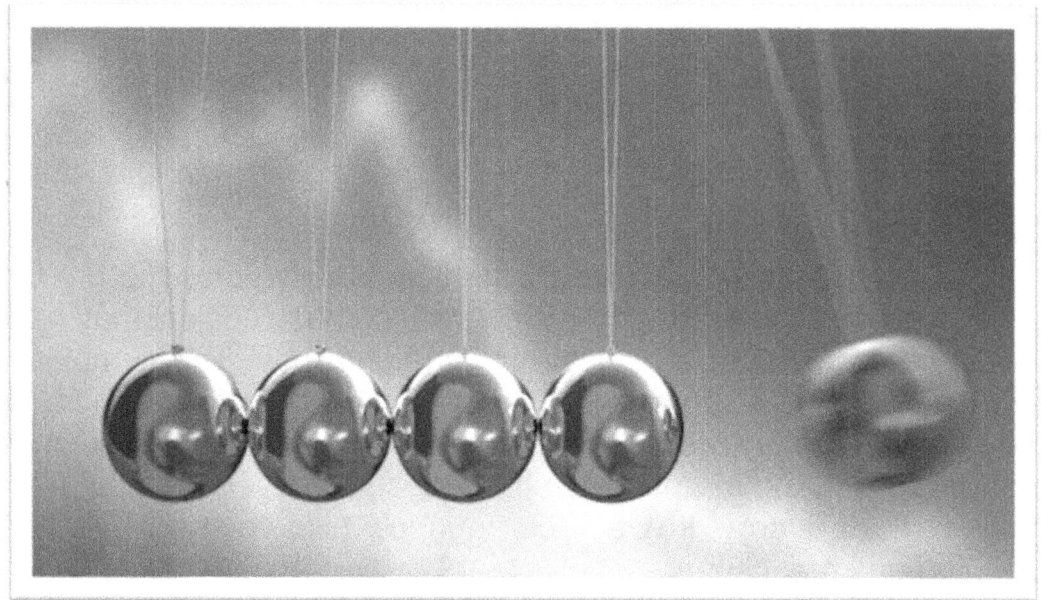

This transition must have come from somewhere, provided that momentum is retained, and the answer to this is that it has been absorbed into the Planet, the momentum of which shifts in the opposite direction by the same amount. However, the velocity change associated with this momentum shift is incredibly small and undetectable in nature since the Planet is enormously more massive than the ball. A collision between two balls, such as on a snooker table, is another example of momentum conservation, where we see how direction, as well as magnitude, are involved in the conservation of momentum.

Temperature

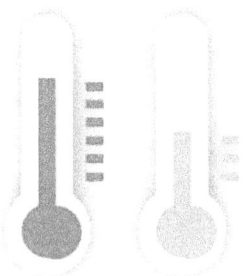

The value of temperature to physics is that it is a measure of heat-related energy. All matter is composed of atoms, as we shall discuss shortly. They are constantly in motion in a gas such as the air in a room and therefore possess kinetic energy. The higher the gas temperature, the higher the average kinetic energy of the gas, and if the gas is cooled to a lower temperature, the molecules will move slower, and the kinetic energy will be lower. We should finally reach a point where the molecules have stopped moving so that the kinetic energy and hence the temperature is zero if we were to continue this process.

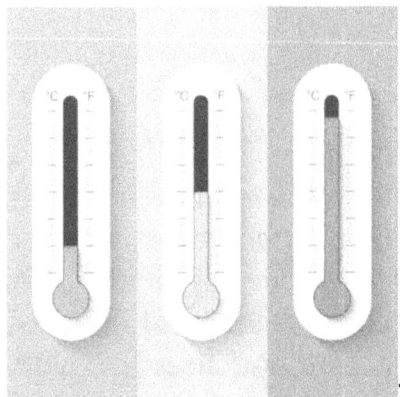

This point is recognized as the 'absolute temperature zero' and on the Celsius scale corresponds to-273 degrees. In solids and liquids, atoms and molecules are both in thermal motion, but the specifics are somewhat different: in solids, for example, the atoms are kept close to and vibrate around specific points. In any case, however, as the temperature is lowered and stops as absolute zero is reached, this thermal motion decreases.

In order to describe an 'absolute degree' of temperature, we use the definition of absolute zero. The degree of this scale's temperature is the same as that of the Celsius scale, except the zero is equal to absolute zero. Temperatures on this scale are known as 'absolute temperatures' or 'kelvins' (abbreviated as 'K'). Thus, absolute zero degrees (i.e., 0 K) corresponds to-273 ° C, while a room temperature of 20 ° C equals 293 K, the water boiling point (100 ° C) is 373 K, and so on.

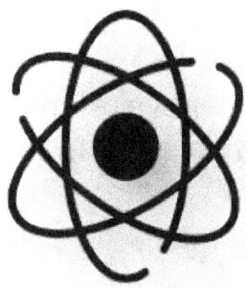

The Quantum Objects

In the latter half of the nineteenth century, the need for radically new physical theories arose as scientists found themselves being unable to account for some of the manifestations that had recently been discovered. Some of these were linked to a thorough analysis of light and similar radiation, to which we will return in the next chapter, whilst others emerged from the study of matter and the discovery that 'atoms' are made of.

Atom

Since the time of the ancient Greek philosophers, there has been speculation that if the matter were divided into smaller and smaller sections, a point would be reached where it was impossible to subdivide further. In the nineteenth century, these theories were established when it was recognized that the characteristics of various chemical elements could be attributed to the fact that they were composed of atoms that were similar but varied from element to element in the case of a particular element.

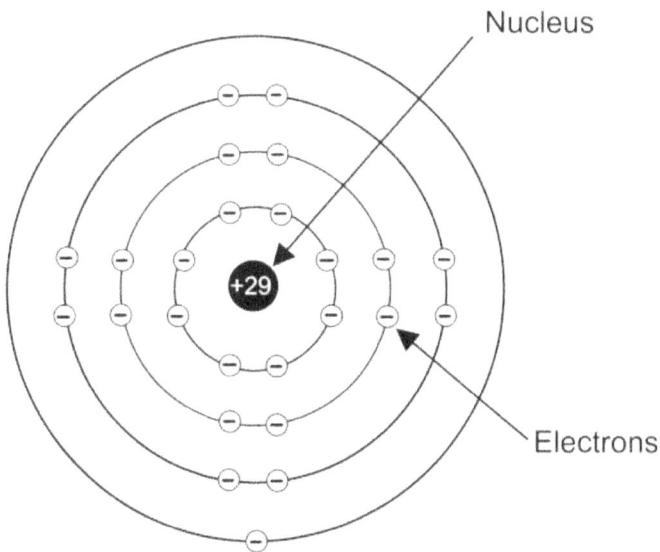

Thus, a hydrogen gas container consists of only one type of atom (known as a hydrogen atom), only another type of carbon lump (i.e., carbon atoms), and so on. It has become possible to measure the size and mass of atoms by various methods, such as studies of the precise properties of gases.

These are very small on the scale of everyday objects, as expected: the size of an atom is about 10-10 m and, in the case of hydrogen, it weighs between about 10-27 kg and, in the case of uranium, 10-24 kg (the heaviest naturally occurring element). While atoms are the smallest objects that bear the identity of a particular element, they are made from a 'nucleus' and many 'electrons' and have an internal structure.

Electron

Electrons are matter particles that weigh much less than the atoms that contain them, with an electron's mass being a little less than 10-30 kg.

They are 'point particles,' which suggests that their size is zero or at least too small to have been determined by any experiments carried out to date. All electrons bear an equal negative electric charge.

Nucleus

Almost all of the atom's mass is contained in a 'nucleus' that is much smaller than the whole atom, usually 1015 m in diameter or around 105 times the atom's diameter. In order to make the atom uncharged or 'neutral' overall, the nucleus bears a positive charge equal and opposite to the total charge borne by the electrons. It is understood that the nucleus, along with some uncharged particles known as 'neutrons, can be further divided into some positively charged particles known as' protons '; the charge on the proton is positive, equal, and opposite to that on the electron.

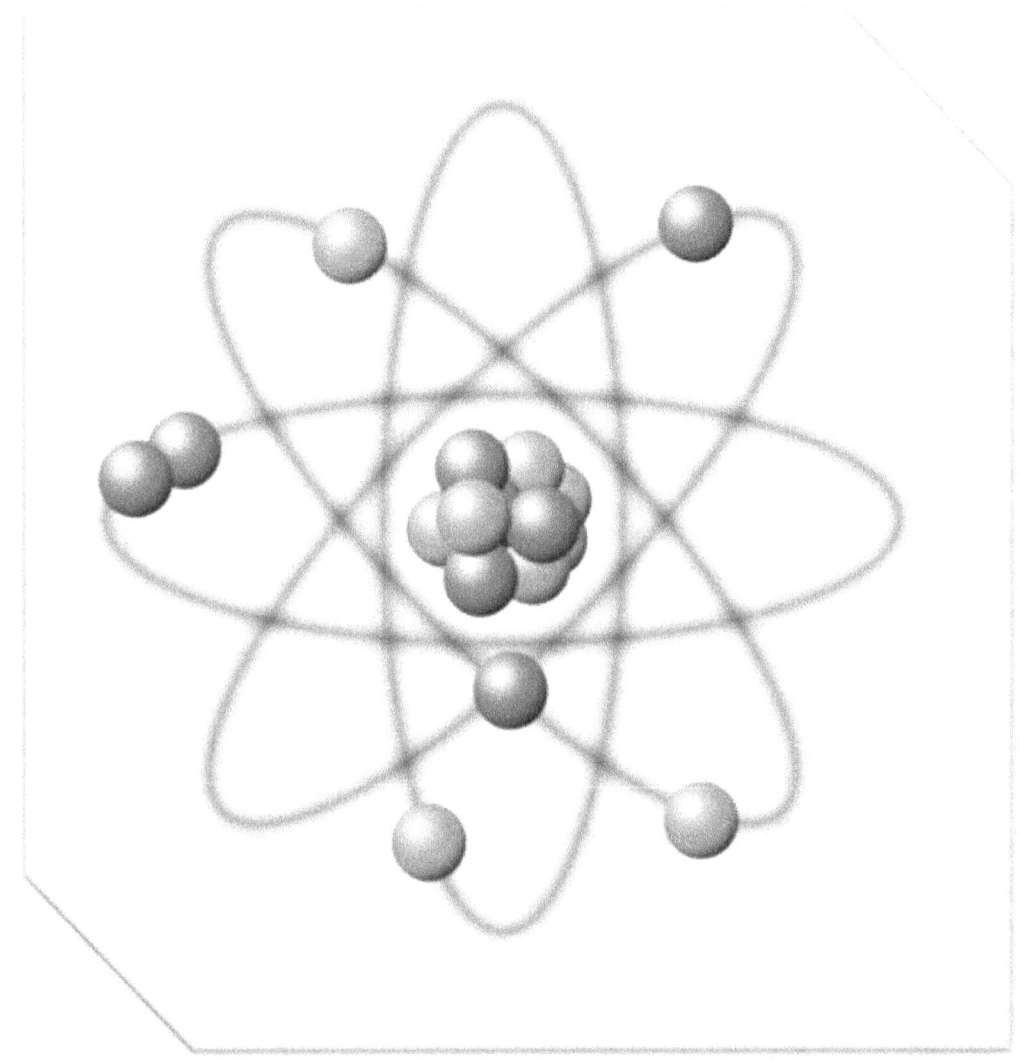

The neutron and proton masses are somewhat similar (though not identical) to each other, both being about two thousand times the mass of the electron. The hydrogen nucleus containing one proton and no neutrons are examples of nuclei; the carbon nucleus containing six protons and six neutrons; and the uranium nucleus containing ninety-two protons and between 142 and 146 neutrons-see 'isotopes' below.

We call it a 'nucleon' when we want to refer to one of the particles making up the nucleus without knowing whether it's a proton or a neutron. Nucleons, like the electron, are not pointed particles but have a structure of their own. They are each made from three-point particles referred to as 'quarks.' In the nucleus, two kinds of quarks are present, and these are known as the 'up' quark and the 'down' quark, but these names should not be correlated with any physical meaning. Up and down quarks bear positive value charges, 2/3 and 1/3 of the overall charge on a proton, which comprises two up and one down quarks, respectively.

The neutron is built from one quark up and two quarks down, which is consistent with its absolute zero charges. In almost all cases, the quarks inside a neutron or proton are bound together very closely so that the nucleons can be viewed as single particles. The neutrons and protons interact less strongly but also interact much more strongly than the electrons, which means that a nucleus can also be viewed as a single particle to a very good approximation, and its internal structure is overlooked when we consider the atom's structure.

Isotopes

The majority of atomic properties are derived from electrons, and the number of electrons charged negatively is equal to the number of protons charged positively in the nucleus. The nucleus, however, also contains several uncharged neutrons, as mentioned above, which contribute to the mass of the nucleus but otherwise do not significantly affect the atom's properties.

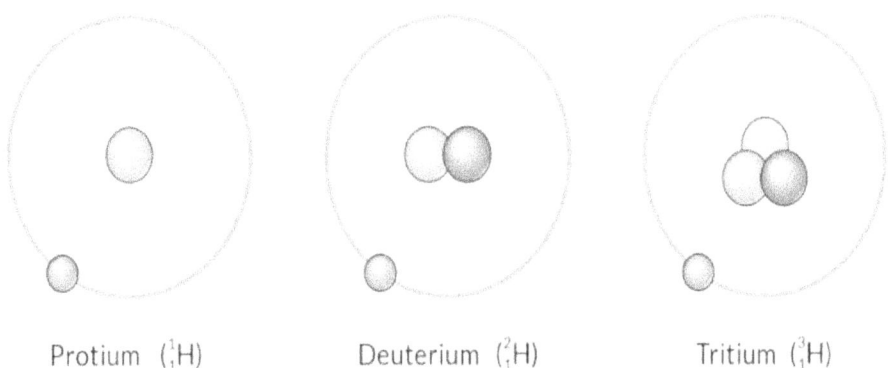

Protium ($_1^1$H)   Deuterium ($_1^2$H)   Tritium ($_1^3$H)

They are classified as 'isotopes' if two or more atoms have the same number of electrons (and hence protons) but different numbers of neutrons. An example is 'deuterium,' whose nucleus comprises one proton and one neutron, and which is thus an isotope of hydrogen; approximately one atom in every ten thousand is deuterium in naturally occurring hydrogen.

The number of isotopes, i.e., those with a higher number of nucleons, varies from element to element and is greater for heavier elements. Uranium, which has nineteen isotopes, all of which has 92 protons, is the strongest naturally occurring element. U238, which comprises 146 neutrons, is the most common of these, while the isotope included in nuclear fission (see Chapter 3) is U235 with 143 neutrons. Note the notation where the total number of nucleons is the superscript number.

Atomic Structure

We have shown so far that an atom consists of a very small nucleus that is positively charged, surrounded by many electrons. The simplest atom is hydrogen, with one electron, and uranium, which comprises ninety-two electrons, is the largest naturally occurring atom. It is obvious that a large part of the volume filled by the atom must be a vacuum, realizing that the nucleus is very small and that the electron's dimensions are essentially zero. This means that, even though there is an electrical attraction between each negatively charged electron and the positively charged nucleus, the electrons must remain some distance from the nucleus.

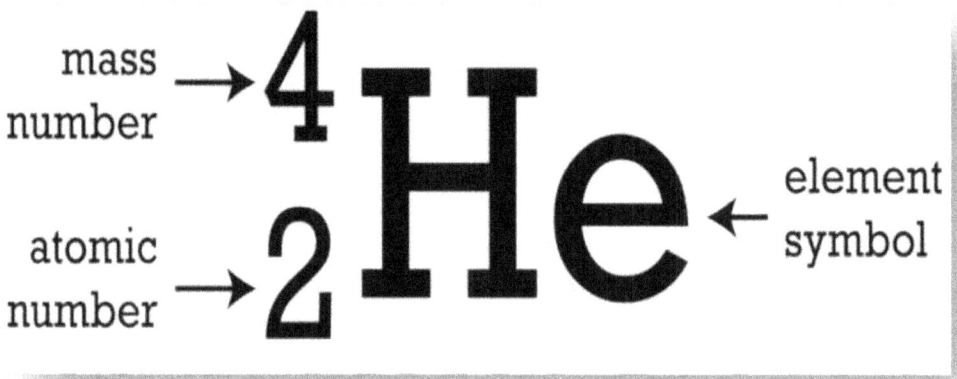

Why doesn't an electron fall into the nucleus, then? One theory proposed early in the subject's development is that the electrons are in orbit around the nucleus, much like the planets in the solar system orbiting the sun. However, a significant difference is that orbital charges are known to lose energy by emitting electromagnetic radiation such as light between satellite orbits in a gravitational field and those where the orbiting particles are charged.

They should travel closer to the nucleus to save energy, where the potential energy is lower, and calculations indicate that this should lead to a small fraction of a second of the electron falling into the nucleus. However, this does not and must not occur in order for the atom to have its known size. This observed property of atoms cannot be accounted for by any model based on classical physics, and a new physics, quantum physics, is needed.

A basic atomic property that is incomprehensible from a classical point of view is that all the atoms associated with a specific element are identical. The atom would have all the properties associated with the product, provided it contains the correct number of electrons and a nucleus bearing a compensating positive charge. Thus, one electron is found in a hydrogen atom, and all hydrogen atoms are equal. Think again about a traditional orbiting dilemma to see if this is classically shocking.

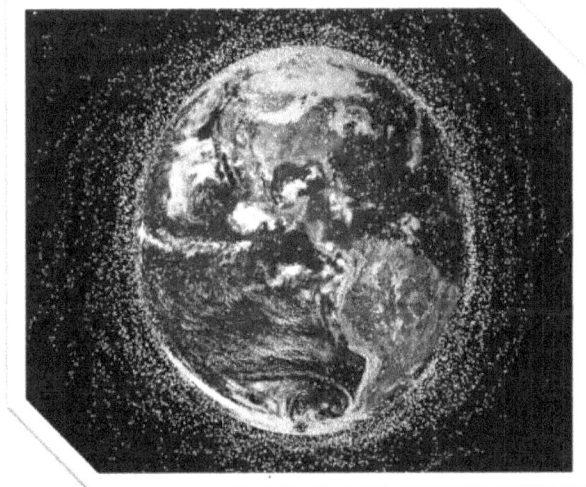

If we place a satellite in orbit around the Earth, then it can be at any distance from the Earth that we want, provided we do rocket science properly. But all hydrogen atoms are the same size, which not only means that their electrons must be kept at a certain distance from the nucleus but also implies that this distance is the same at all times for all hydrogen atoms (unless an atom is intentionally 'excited' as we discuss below). Once again, we see that the atom has properties that are not explainable.

Consider what we would do to an atom to alter its size to explore this argument further. We will have to inject energy into the atom as pushing the electron away from the nucleus increases its electrical potential energy, which has to come from somewhere. This can be done without getting too deep into the functional specifics by moving an electric discharge through a gas consisting of atoms. We notice the energy is naturally absorbed and then re-emitted in the form of light or other sources of electromagnetic radiation.

If we do this: we see this happening if a fluorescent light is turned on. It seems that it returns to its initial state by releasing radiation when we excite the atom in this manner, rather than as we expected in the case of a charge in a classical orbit.

Atomic Radiation

There are, however, two major variations in the case of atoms. The first, discussed above, is that for all atoms of the same form, the final configuration of the atom corresponds to the electron being some distance from the nucleus, and this state is always the same. The second distinction has to do with the existence of the released radiation.

Radiation is in the form of electromagnetic waves, which will be explored in more detail in the next chapter; we only need to know for the moment that such a wave has a characteristic wavelength corresponding to the light color. Classically, the light of all colors should be produced by a spiraling charge, but when the light emitted by an atomic discharge is analyzed, it is found to contain only certain colors matching unique wavelengths.

These form a fairly simple pattern in the case of hydrogen, and it was one of the key early triumphs of quantum physics that it was able to predict this quite accurately. The principle that the potential values of an atom's energy are limited to such 'quantized' values, which include the lowest value or 'ground state' in which the electron stays some distance from the nucleus, is one of the latest ideas on which this is based. As the atom consumes energy, it will only do so if one of the other permitted values ends up with the energy. The atom is said to be in an 'excited state' with the electron further from the nucleus than it is in the ground state. It then returns to its ground state, releasing radiation, the wavelength of which is determined by the energy difference between the initial and final states.

CHAPTER 2: WAVES AND PARTICLES

Many people have heard that a major aspect of quantum mechanics is 'wave-particle duality.' We will try to explain what this means in this chapter and how it allows us to understand a number of physical phenomena, including the atomic structure issue that I presented at the end of the previous chapter. We can find that the effects of certain physical processes are not precisely calculated at the quantum level, and the most we can do is estimate the likelihood of 'probability' of different future events. In evaluating these probabilities, we will find that something called the 'wave function' plays an important role: its power, or intensity, for instance, at any point, represents the likelihood that we will detect a particle at or near that point.

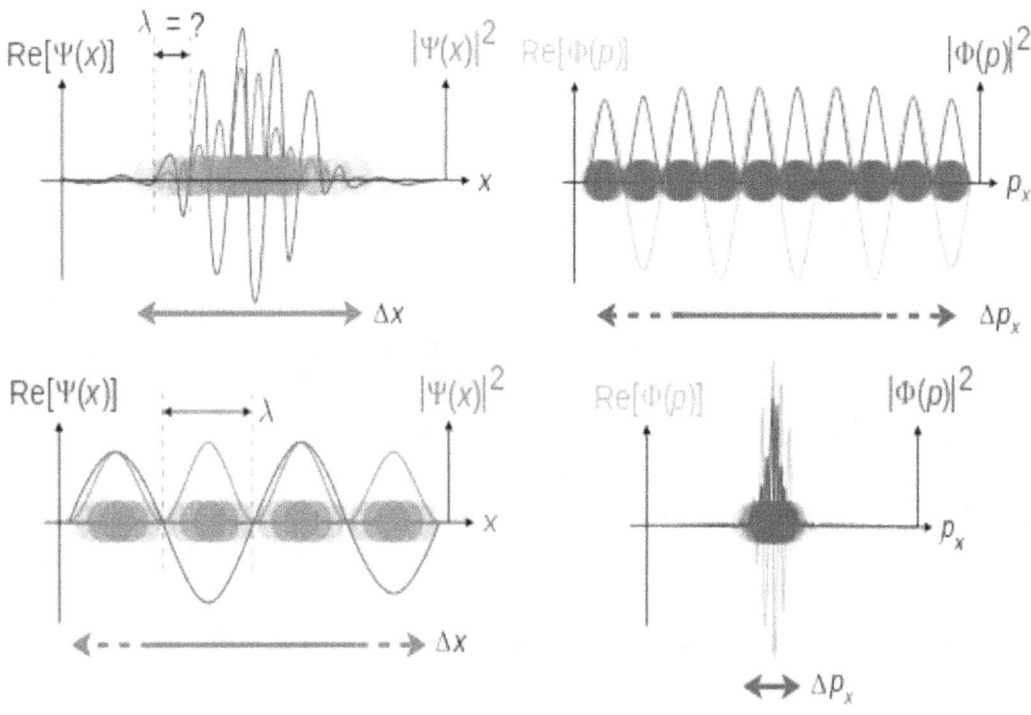

We have to know more about the wave function relevant to the physical situation we are considering in order to make progress. By solving a very complex mathematical equation, known as the Schrödinger equation (after the Austrian physicist Erwin Schrödinger, who discovered this equation in the 1920s), trained quantum physicists calculate it; but without doing this, we can find that we can get quite a long way. Instead, we're going to build up an image based on some of the fundamental characteristics of waves, and we're going to start with a discussion of them as they appear in classical physics. We all have a certain experience with waves.

Ocean waves would be known to those who have lived near or visited the seacoast or have traveled on a ship. They can be very big, impacting ships with violent results, and when they roll on a beach, they provide surfers with entertainment. However, it would be more helpful for our purposes to think of the more gentle waves or ripples that occur when an object is dropped into a calm pond, such as a stone.

That shows a profile of such a wave, showing how it shifts at various locations in time. The water surface oscillates up and down in a normal way at every specific point in space. The ripple height is known as the 'amplitude' of the wave, and the 'period' is known as the time taken for a complete oscillation. It is also beneficial to refer to the wave's 'frequency,' which is the number of times a second it passes over a full oscillation period. The form of the wave repeats in space at any point in time, and the repeat distance is known as the 'wavelength.' The pattern travels over a distance equal to the wavelength during a time corresponding to one duration, which implies that the wave moves at a speed corresponding to one wavelength per period.

Traveling Waves and Standing Waves

Since they 'travel' in space, waves are what are called 'traveling waves.' The motion is from left to right in the illustration shown, but it could also have been from right to left indeed that the ripples spread out in all directions from a stone dropped in water. We will have to hear about 'standing waves' as well as moving waves.

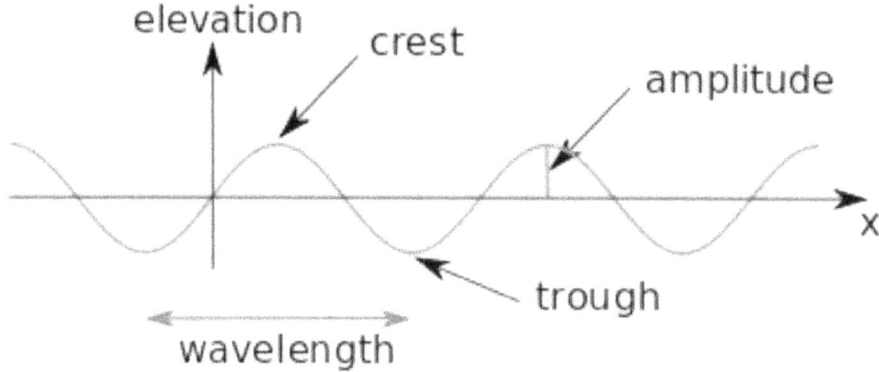

An example, we see that the wave has a shape similar to that previously mentioned, and the water oscillates up and down again, except now the wave does not travel along but remains in the same position, hence its name. When it is confined to a 'cavity' surrounded by two borders, a standing wave usually occurs. It is mirrored at one of the boundaries and travels back in the opposite direction if a moving wave is set up. The net effect is the standing wave when the waves moving in the two directions are combined. In certain cases, the cavity walls are such that they can not be penetrated by the wave, and this results in the amplitude of the wave being equal to zero at the borders of the cavity.

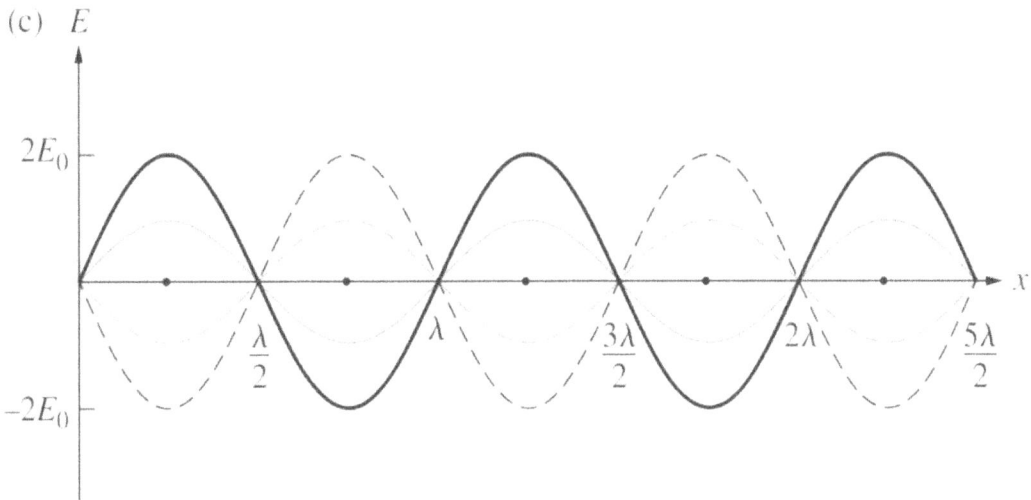

(c) E

$2E_0$

$-2E_0$

$\frac{\lambda}{2}$ λ $\frac{3\lambda}{2}$ 2λ $\frac{5\lambda}{2}$ x

This means that only standing waves of unique wavelengths will fit into the cavity, so for the wave to be zero at both borders, for a whole range of peaks or troughs to fit into the cavity, the wavelength must be just the right length.

However, the sound would never hit our ears if the standing waves were the whole story. The instrument's vibrations must produce moving waves in the air, which bring the sound to the listener for the sound to be transmitted to the listener. The body of the instrument oscillates in sympathy with the string in a violin, for example, and creates a moving wave that radiates out to the audience.

Most of the science (or art) of designing musical instruments consists of ensuring that in the emitted moving waves, the frequencies of the notes identified by the permissible wavelengths of the standing waves are reproduced.

A complete understanding of the actions of musical instruments and how they convey sound to an audience is a significant subject in itself, which we do not need to go deeper into here. A book on the physics of music should be consulted by interested readers.

Electromagnetic radiation, exemplified in the electromagnetic waves that carry signals to our radios and televisions and in light, is another widely observed wave-like phenomenon. These waves have different frequencies and wavelengths: standard FM radio signals, for example, have a wavelength of 3 m, while the wavelength of the light depends on the color, with blue light being approximately 4 x 108 m and red light being approximately 7 x 10-8 m; other colors have wavelengths between those values.

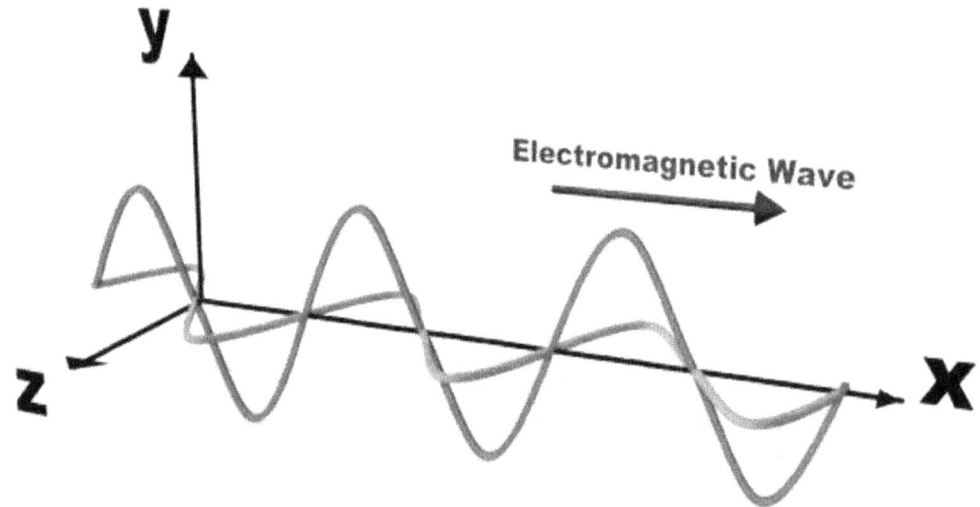

Light waves differ from water waves and sound waves in that, in the examples discussed earlier, nothing corresponds to the vibrating medium (i.e., water, string, or air). Indeed, as is evident from the fact that we can see the light produced by the sun and stars, light waves are capable of traveling across space. In the eighteenth and nineteenth centuries, this property of light waves posed a major issue to scientists.

Some concluded that space is not empty but filled with an otherwise undetectable material known as 'aether' that was believed to support lightwave oscillation. However, when it was discovered that the properties needed to accommodate the very high frequencies typical of light could not be reconciled with the fact that the aether does not provide any resistance to the passage of objects (such as the Earth in its orbit) through it, this theory ran into trouble.

Interference

Direct proof is derived by studying 'interference' that a phenomenon, such as light, is a wave. When two waves of the same wavelength are added together, interference is usually observed. We see that if the two waves are in step ('in phase' is the technical term), they join together to create a combined wave twice the amplitude of each of the originals. If, on the other hand, they are precisely out of step, they cancel each other out (in 'antiphase'). The waves partly cancel in intermediate situations, and there is a value between these extremes of the combined amplitude. Interference is an important proof of the wave properties of light, and this effect can not be accounted for by any other classical model. For instance, suppose we had two classical particle streams instead: the total number of particles would always be equal to the sum of the numbers in the two beams, and they would never be able to cancel each other out in the way waves do.

WAVE INTERFERENCE

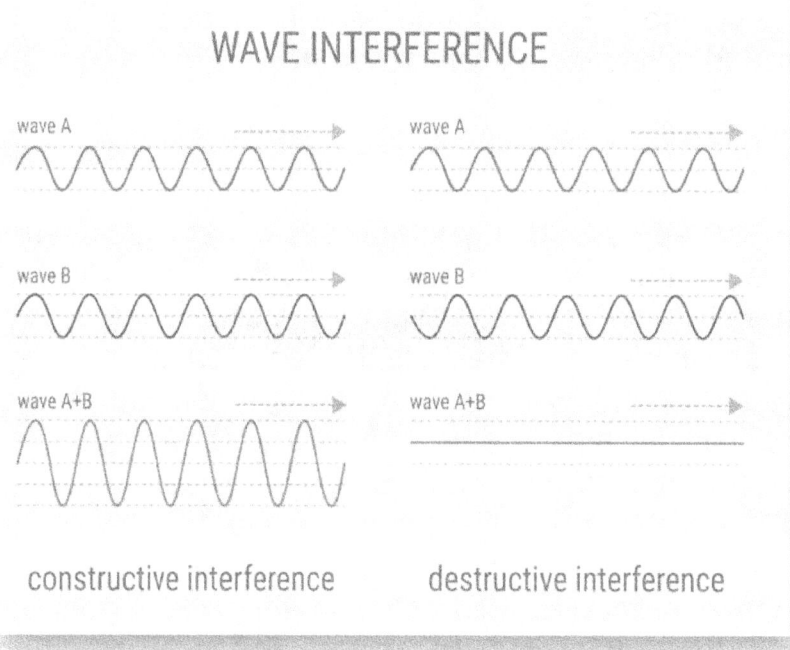

wave A

wave B

wave A+B

constructive interference

wave A

wave B

wave A+B

destructive interference

Thomas Young, who experimented around 1800, was the first person to observe and describe interference (c). Light passes through an O-labeled narrow slit, after which it meets a two-slit screen, A and B, and eventually enters the third screen, S, where it is observed. Either of two routes could have traveled by the light hitting the last panel, either by A or by B. The distances traveled by the light waves following these two paths, however, are not identical, so they normally do not arrive in step with each other on the frame. It follows from the discussion in the previous paragraph that the weaves will reinforce each other at certain points on S, while at others, they will cancel; as a result, a pattern is observed on the screen consisting of a sequence of light and dark bands.

In some cases, it exhibits particle properties, and a fuller understanding of the quantum essence of light will introduce us to 'wave-particle duality.'

Evidence started to appear towards the end of the nineteenth century and the beginning of the twentieth, indicating that it is not sufficient to classify light as a wave to account for all of its observed properties. Two separate research areas were central to this. This heat radiation becomes noticeable at relatively high temperatures, and we identify the object as 'red hot' or, at even higher temperatures,' gives off a white heat.'

We note that red corresponds to the longest wavelength in the optical spectrum, so it appears that it is easier to produce long-wavelength light (i.e., at a lower temperature) than shorter wavelength light; indeed, long-wavelength heat radiation is generally referred to as 'infrasound.'

Physicists sought to understand the properties of heat radiation following the advent of Maxwell's theory of electromagnetic radiation and improvement in the understanding of heat (a field to which Maxwell himself made significant contributions).

It was then understood that temperature is energy-related: the hotter an object is, the more energy it contains from heat.

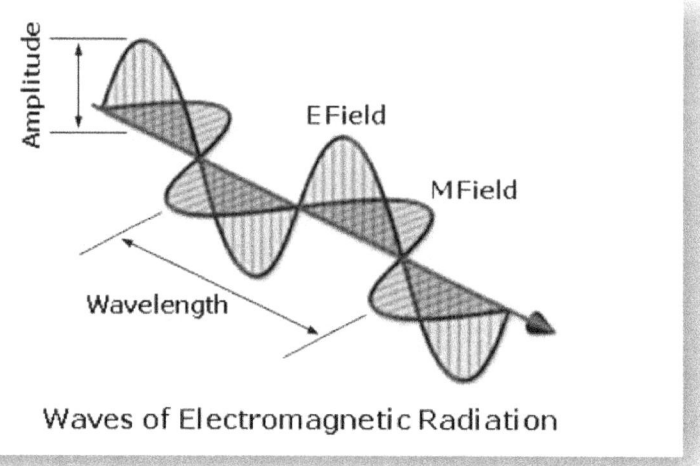

Amplitude

E Field

M Field

Wavelength

Waves of Electromagnetic Radiation

Also, Maxwell's theory predicted that an electromagnetic wave's energy should depend only on its amplitude and should be independent of its wavelength in particular. Therefore, as the temperature increases, one would imagine that a hot body will radiate at all wavelengths, the radiation being brighter but not changing color.

Detailed calculations showed that since the number of potential waves of a given wavelength increases as the wavelength decreases, the heat radiation of the shorter wavelength should actually be brighter than that of long wavelengths, but at all temperatures, it should be the same again. All objects could appear violet in color if this were valid, their average brightness being low at low temperatures and high at high temperatures, which is not what we observe, of course. This disparity was known as the 'ultraviolet catastrophe' between theory and observation.

Matter Waves

The fact that light has particle properties, and is conventionally called a wave, led the French physicist Louis de Broglie to speculate that other artifacts that we normally think of as particles may have wave properties. Thus, in certain cases, a beam of electrons, which is most naturally conceived as a stream of very small bullet-like particles, will act as if it were a wave. Davidson and Germer first explicitly confirmed this radical concept in the 1920s: they passed an electron beam through a graphite crystal and found a pattern of interference that was similar in principle to that created when light passes through a series of slits.

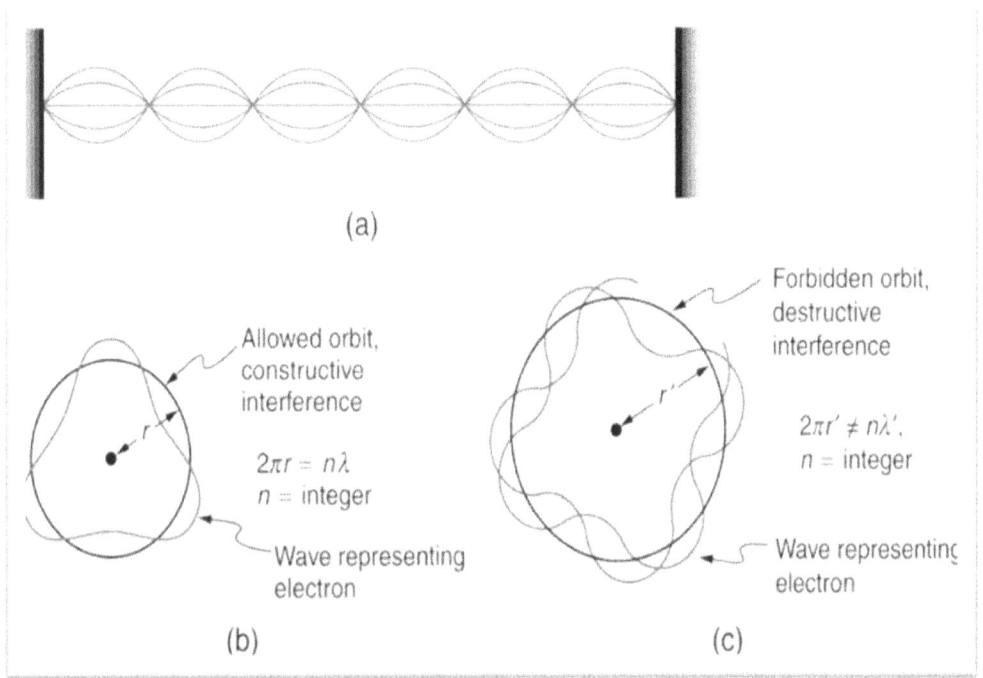

(a)

Allowed orbit, constructive interference

$2\pi r = n\lambda$
$n =$ integer

Wave representing electron

(b)

Forbidden orbit, destructive interference

$2\pi r' \neq n\lambda'$,
$n =$ integer

Wave representing electron

(c)

As we have shown, this property is fundamental to the proof that a wave is light, so this experiment is a clear confirmation that electrons can also be added to this model. Similar evidence was later found for the wave properties of heavier particles, such as neutrons, and wave-particle duality is now assumed to be a universal property of all particle forms. Even ordinary objects such as grains of sand, footballs, or motorcars have wave properties, but in these cases, the waves are totally unobservable in nature – partially because the appropriate wavelength is far too small to be visible, but also because classical objects are composed of atoms, each of which has its associated wave and all these waves are constantly chopping and shifting.

We have shown above that in the case of light, the wave vibration frequency is directly proportional to the quantum energy.

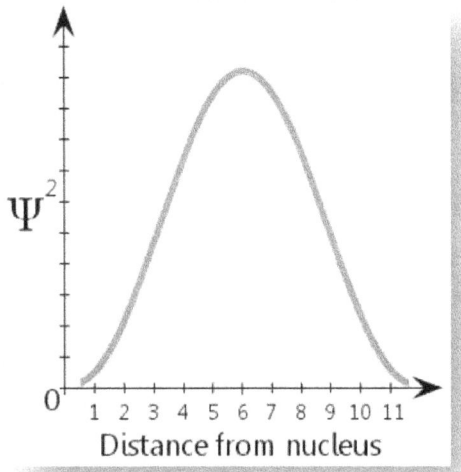

The frequency turns out to be hard to describe in the case of matter waves and difficult to calculate directly. There is, instead, a relation between the wavelength of the wave and the momentum of the object, so that the higher the momentum of the particle, the shorter the wavelength of the wave of matter. The surface of the water goes up and down, the air pressure oscillates in sound waves, and in electromagnetic waves, electric and magnetic fields differ.

What is the corresponding number in the case of waves of matter? The conventional answer to this question is that this corresponds to no physical quantity. We can do the wave estimation,

Using quantum mechanics principles and equations, we can use our results to estimate the values of quantities that can be experimentally tested, but we can not detect the wave itself directly, so we do not need to describe it physically and do not attempt to do so. We use the term 'wave function' rather than wave to emphasize this, which illustrates the fact that it is a mathematical function rather than a physical entity.

Another important technical distinction between wave functions and the classical waves we discussed earlier is that, while the classical wave oscillates at the wave frequency, the wave function stays constant in time in the matter-wave case. However, while not physical in itself, the wave function plays an integral role in the application of quantum mechanics to the understanding of waves.

First, if the electron is confined to a given area, the wave function forms standing waves similar to those described earlier; as a consequence, one of a collection of discrete quantized values is taken up by the wavelength and thus the momentum of the particle. Second, if we conduct experiments to detect the presence of the electron near a specific point, we are more likely to find it in regions where the function of the wave is high than in those where it is small.

Max Born, whose rule states that the likelihood of finding the particle near a certain point is proportional to the square of the magnitude of the wave function at that point. Atoms contain electrons that are restricted to a small region of space through the electrical force attracting them to the nucleus. We may expect the related wave functions to form a standing-wave pattern from what we said earlier, and we can soon see how this leads to an understanding of essential atomic properties. We start this debate by looking at a simpler framework in which we imagine that an electron is contained inside a small box.

Electron in a Box

We consider the case of a particle in this instance, which we will assume to be an electron stuck inside a box. By this, we say that if there is an electron in the box, there is a constant value of its potential energy, which can be taken to be zero. The electron is confined to the box since it is surrounded by a region of very high potential energy, which, without violating the principle of conservation of energy, the electron will not reach.

A ball inside a square box sitting on the floor would be a classical analogy: if the sides of the box are high enough, the ball will not escape from the box, so it would need to overcome gravity to do so. We will soon consider the material waves suitable to this situation, and we might compare them with the case of a pond or swimming pool, where the water is surrounded by a solid border: the solid shore is unable to vibrate, so the water must be confined to any waves produced.

We regard the issue as 'one-dimensional' as a further simplification, meaning that the electron is limited to moving in space in a specific direction such that motion in the other directions can be ignored. On a line, we can then make an analogy with waves, which are actually one-dimensional since they can only travel down the string. Now we are considering the shape of the function of the electron wave. Since the electron does not escape from the box, there is zero possibility of finding it outside.

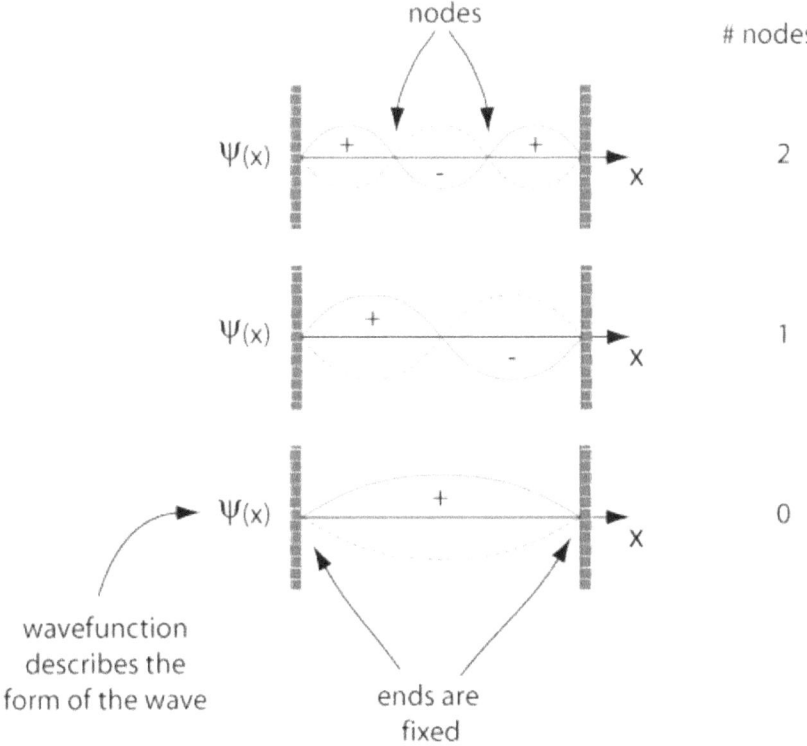

The chance of finding the particle at that point can only have one value if we consider the very edge of the box, so the fact that it is zero outside the box means that just inside it must also be zero. This situation is very close to that of a violin or guitar string, and we saw earlier that this means that the wave must be a standing wave with a wavelength that fits into the available space.

This is seen in the above example, and we see that one of the values corresponding to a whole number of half wavelengths fitting into the box is limited to the wavelength of the wave. This means that only these unique wavelength values are permitted, and since the electron momentum is determined by the wavelength via the de Broglie relationship, the momentum is therefore limited to a specific set of values.

Therefore, if we had several similar electron-containing boxes, their soil states would also be identical. One of the characteristics of atoms that we could not classically describe was that all atoms of a given form have the same characteristics, and that they all have the same lowest energy state, in particular. Quantum physics has demonstrated, by wave-particle duality, why such a condition occurs in the case of an electron in a box, and we shall soon see how the same concepts apply to an electron in an atom.

Varying Potential Energy

The matter waves associated with particles propagating in free space or captured in a one-dimensional box have been considered so far. The particle travels in an area where the potential energy is constant in all these situations because if we remember that the total energy is conserved, the kinetic energy and thus the momentum and speed of the particle would be the same everywhere it goes. In comparison, for example, a ball rolling up a hill absorbs potential energy, loses kinetic energy, and as it climbs, it slows down. We also know that the de Broglie relationship relates the velocity of the particle to the wavelength of the wave, so if the velocity remains constant, this number will also be the same everywhere, which is what we have implicitly assumed.

However, the wavelength must also differ if the speed is not constant, and the wave will not have the reasonably simple form we have considered so far. Therefore, as a particle travels through a region where the potential energy changes, it will also change its speed and therefore, the wave function's wavelength.

Quantum Tunneling

The case of a particle reaching a 'potential phase' we first consider. By this, we mean that at a given stage, the potential increases unexpectedly. In particular, we are interested in the case where the energy of the approaching particle is smaller than the height of the step, so we can expect the particle to bounce back from a classical point of view as soon as it hits the step and then travel backward at the same speed. When we apply quantum mechanics, almost the same thing occurs, but as we shall see, there are significant variations.

Second, we consider the shape of the wave-matter. Based on our earlier discussion, we expect particles entering the phase to be represented by traveling waves moving from left to right, while after they bounce back, the wave will be traveling from right to left. In general, at any given moment, we do not know what the particle is doing, so the wave function to the left of the phase would be a mixture of these, and this is proven when mathematically solving the Schrödinger equation.

In the form of the wave to the right of a phase, what is of real interest? There is no chance of finding the particle there classically, so we would expect the wave function in this area to be zero. However, when we solve the Schrödinger equation, we find that the measured wave function does not become zero until some way to the right of the phase.

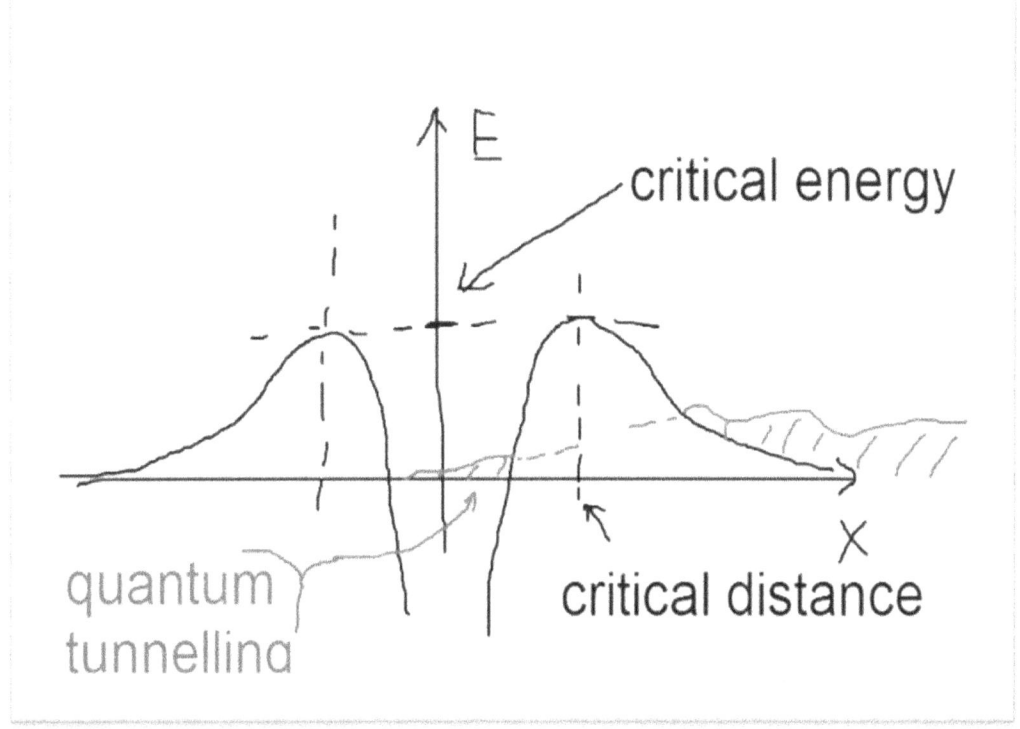

We see that quantum physics predicts that there is a finite probability of finding it in a region where it could never be if classical physics were the entire story, realizing that the amplitude of the wave function at any point reflects the possibility of finding a particle at that point.

A Quantum Oscillator

The second example we consider is the movement of a particle in a parabolic potential. In the classical case, the particle will oscillate periodically from one side of the potential well to the other with a frequency dictated by its mass and shape. The size of the oscillation, or 'amplitude,' is determined by the energy of the particle: all of this energy is kinetic at the foot of the well, while the particle comes to rest at the limits of its motion, where all of the energy is potential. By solving the Schrödinger equation, the wave functions are obtained, and it is found that standing-wave solutions are only possible for unique values of energy.

The Hydrogen Atom

The simplest atom is that of the hydrogen element, which consists of a single negatively charged electron bound by the electrostatic (or 'Coulomb') force to a positively charged nucleus, which is heavy when the electron is close to the nucleus and decreases in intensity gradually when the electron is farther away. As a consequence, the potential energy near the nucleus is high and negative and gets closer to zero as we step away from it.

All the examples discussed so far have been onedimensional, which implies that we have implicitly concluded that the particle is forced to travel in a certain direction (from left to right or vice versa in our diagrams). However, atoms are three-dimensional objects, and before we can grasp them entirely, we will have to take this into account. A significant simplifying characteristic of the hydrogen atom is that 'spherically symmetric' is the Coulomb potential, i.e., it relies only on the distance between the electron and the nucleus, regardless of the direction of this separation. As a result, many of the wave functions associated with the energy levels permitted have the same symmetry; we will first discuss these and then return to the others.

Other Atoms

More than one electron is found in atoms other than hydrogen, creating further complications. Before we can answer these, after its inventor Wolfgang Pauli, we have to consider another quantum theory, known as the 'Pauli exclusion principle.' This states that any specific quantum state, such as an electron, does not contain more than one particle of a given kind. While easily stated, this theory can only be proven through the use of very advanced quantum analysis, and we will definitely not attempt to do this here.

However, we have to know about a further property possessed by quantum particles and known as 'spin' before we can correctly apply the exclusion principle. We know that the planet spins on its axis as it travels around the sun in orbit, so we would well expect the electron to spin similarly if the atom were a classical entity. To some extent, this analogy holds, but once again, there are major variations between classical and quantum circumstances. The spin properties are governed by two quantum rules: first, the rate of spin is always the same for any given form of the particle (electron, proton, neutron); and, second, the direction of spin around any axis is either clockwise or anticlockwise.

This suggests that an electron can have one of only two spin states in an atom. Thus, when they spin in opposite directions, any quantum state represented by a standing wave can contain two electrons. Consider what happens if we put several electrons in the box discussed earlier as an example of the application of the Pauli exclusion principle. All the electrons must occupy the lowest possible energy levels to form the state with the lowest total energy.

Therefore, if we think of adding them one at a time to the box, the first goes into the ground state, as does the second with the opposite spin to the first. This level is now complete, so the third electron, along with the fourth and the spins of adding up to two electrons to each energy state before all are accommodated, must go into the next highest energy level. We are now applying this approach to atoms, considering helium first, which has two electrons. Suppose the fact that electrons exert a repulsive electrostatic force on each other is initially overlooked. In that case, we can measure the quantum states in the same way as we did for hydrogen but allowing the nuclear charge to be doubled.

This doubling means that all energy levels are considerably decreased (i.e., made more negative), but otherwise, the collection of standing waves is very close to those in hydrogen, and when the interactions between the electrons are included, it turns out that this pattern is not greatly altered. Therefore, with an opposite spin in the lowest, the lowest energy state has both electrons. Two of these would be in the lowest state in the case of lithium with three electrons, while the third would be in the next higher energy state.

A total of six electrons can be found in the above configuration: two of these occupy a spherical symmetry state, while the others fill three different non-spherical states. A group of states with the same value of n is known as a 'shell,' and it is called a 'closed-shell' if electrons occupy all these states. Therefore, lithium, like sodium with eleven electrons, has one electron outside a closed shell, i.e., two in the $n = 1$ closed shell, eight in the $n = 2$ closed shell, and one electron in the $n = 3$ shell. It is understood that many of the properties of sodium are similar to those of lithium, and what is known as the 'periodic table' of the elements is based on similar correspondences between the properties of other elements.

In terms of the atomic shell configuration, which in turn is a result of the properties of the quantum waves associated with the electrons, the entire structure of the periodic table can be understood.

CHAPTER 3: THE POWER OF QUANTUM

In this chapter, we will discover how quantum ideas play an essential role in the physics of energy production. Quantum physics was actively interested in energy production as soon as humans discovered fire and how to use it. This is still true for many of the types of energy generation that play a central role in modern life. In our cars, we burn fuel and use gas or oil to heat our houses. In the form of electricity, much of the power we use enter our houses, but it is important to note that this is not a source of power in itself but rather a means of moving energy from one location to another.

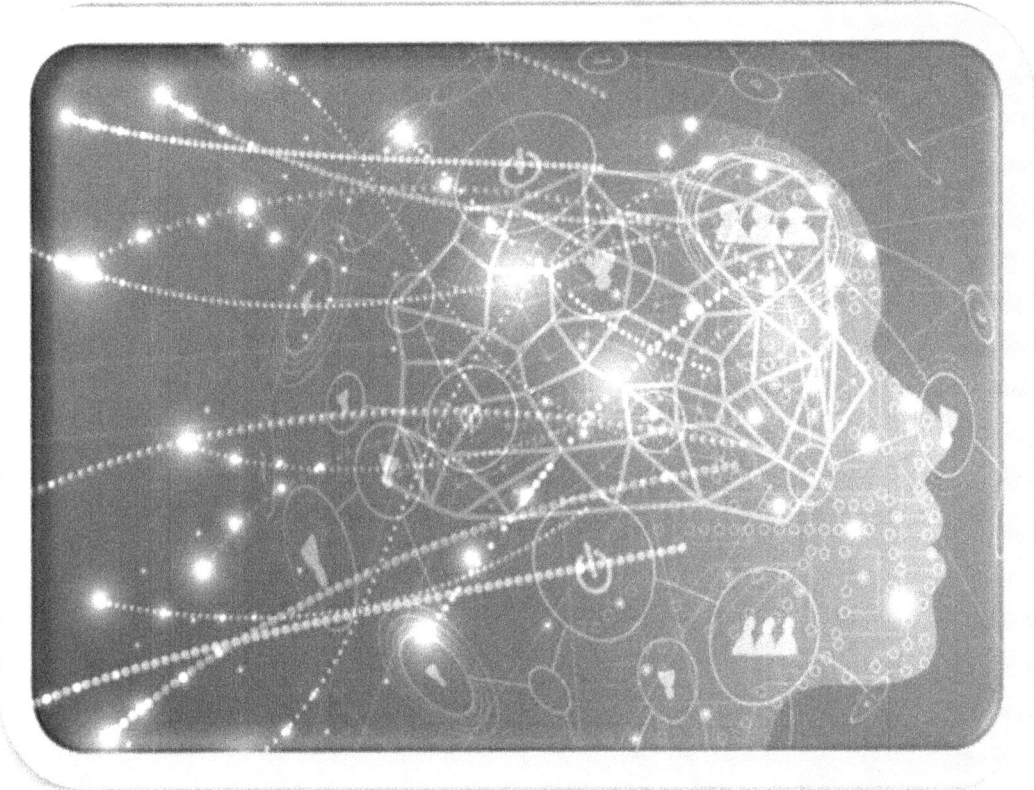

At a power station, electricity is produced from the energy contained in its fuel, which can be a 'fossil fuel' such as coal, oil, or gas; a nuclear fuel such as uranium or plutonium; or a 'sustainable' energy source such as solar, wind or wave energy. Of all of these, only wind and wave power do not depend on quantum physics directly.

Chemical Fuels

Many hydrocarbons contain a fuel such as wood, paper, oil, or coal, which are compounds composed primarily of hydrogen and carbon atoms. As they are heated in the air, hydrogen, and carbon combine to create water and carbon dioxide with oxygen from the air. Energy is produced in the process in the form of heat, which can then be used, for example, to generate electricity in a power station or to power a motor vehicle. We begin with the simplest instance of chemical combination, which is two hydrogen atoms coming together to form a hydrogen molecule, to see how this depends on quantum physics.

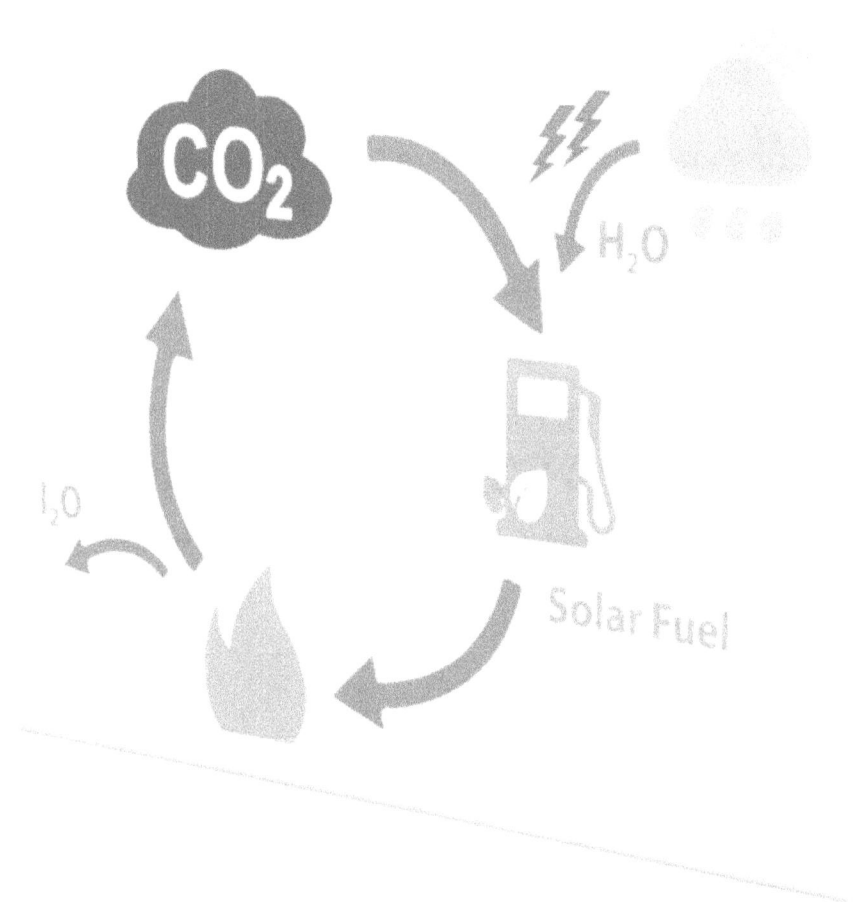

The hydrogen atom contains one proton-attracted electron whose charge is equal to and opposite to that of the electron, so two protons and two electrons are made of a hydrogen molecule. In Chapter 2, we demonstrated how the electron's wave properties result in the hydrogen atom's energy being quantified such that its value is equal to one of a number of unique energy levels; the atom is in the lowest of these energy states, known as the 'earth state' in the absence of excitation.

If we bring two hydrogen atoms into each other, now imagine how the total energy of the system would be affected. The potential energy that changes in three ways is considered first. First, it increases because of the electrostatic repulsion between the two protons that are positively charged; second, it decreases because both protons are now subject to attraction by each electron; third, it increases because of the repulsion between the two electrons that are negatively charged. Also, since the electrons will pass around and between the two nuclei, the kinetic energy of the electrons decreases, so the size of the effective box confining them is increased.

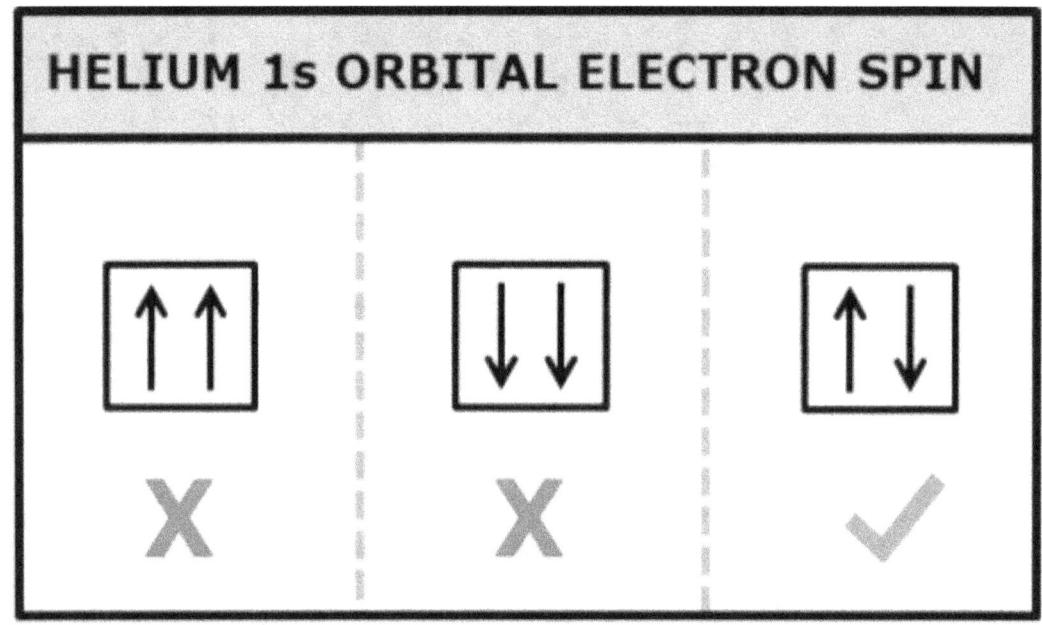

We note that Pauli's exclusion theory allows all electrons to occupy the ground state if they have the opposite spin. The net result of all these changes depends on how far the atoms are apart: there is no difference in total energy when they are wide apart, and electrostatic repulsion between the nuclei dominates when they are very similar. However, there is an overall decrease in the total energy at intermediate distances, and this decrease is at its highest when the protons are around 7. 4 Roughly 10-10 m apart.

At this point, approximately one-third of the hydrogen atom's ground-state energy is equal to the difference between the molecule's energy and that of the widely separated hydrogen atoms. Where does it end up with this surplus energy? The response is that some of it go into the moving molecule's kinetic energy, while the rest is given off in the form of photons. Both are efficient sources of heat, so an increase in temperature, which is just what we expect from fuel, is the ultimate result.

As we shall see later, both useful chemical fuels and indeed nuclear energy underlie the concepts involved in this case. A hydrocarbon fuel, such as oil or gas, comprises molecules that have remained stable for a long time, perhaps millions of years, consisting primarily of carbon and hydrogen. Even when the substances are exposed to air at room temperature, this equilibrium is retained, but if energy is supplied to separate the molecules, the atoms rearrange themselves with the release of energy into a mixture of water and carbon dioxide.

The concepts involved are those of quantum physics: the total energy of the water and carbon dioxide molecules' quantum ground states is smaller than that of the original hydrocarbon molecules. However, to induce this transition, energy must be supplied; once the mixture has been heated to a sufficiently high temperature, the process becomes self-sustaining, and energy continues to be released until the fuel is depleted (unless the process is extinguished).

Nuclear Fuels

Nuclear power concepts are remarkably similar to those underlying the combustion of chemical fuels, but the amount of energy involved in nuclear processes is much larger. As we saw in Chapter 1, the nucleus of an atom is made up of many protons and neutrons bound together by a strong nuclear force.

Even though the specifics are much more complicated than the atomic case mentioned in Chapter 2, the structure of the nucleus is still subject to the laws of quantum physics.

This is because the latter is regulated by the electrons' attraction to the heavier nucleus, while the interactions inside the nucleus between the protons and neutrons are all of a similar mass. In both cases, however, the results are very similar: the energy of the nucleus is quantized into a number of energy levels, like the atom, the lowest of which is known as the 'ground state.'

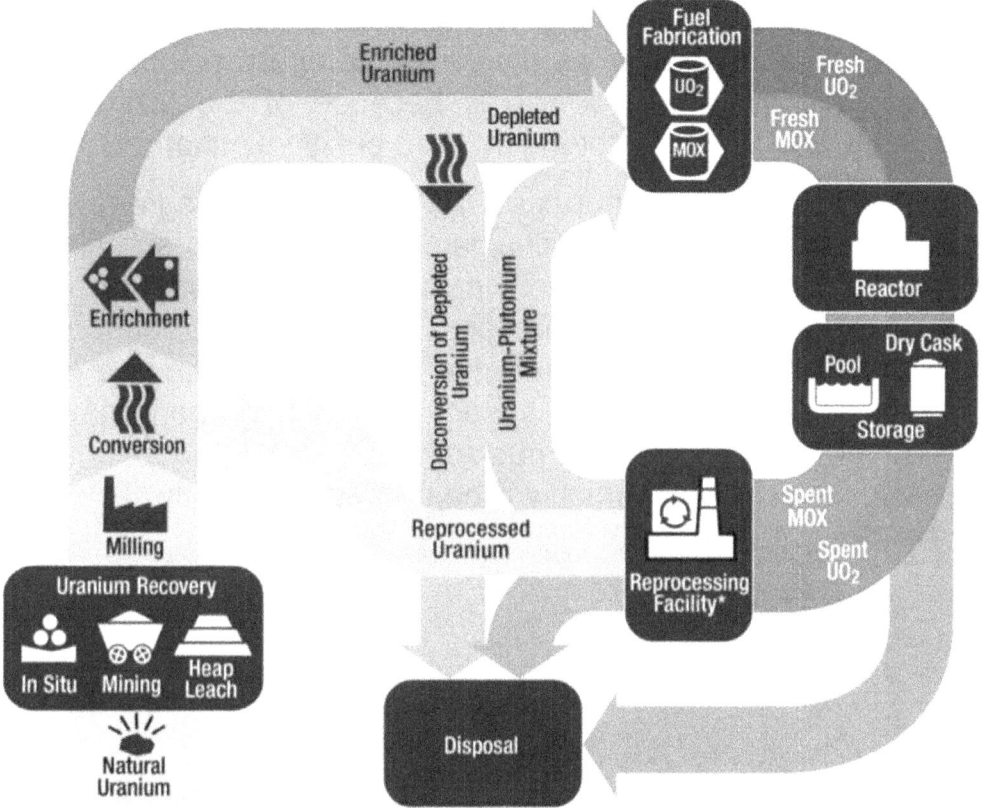

The 'fusion' of two hydrogen nuclei into a single nucleus is closely similar to the earlier example of the merger of two hydrogen atoms into a hydrogen molecule. The nuclei of hydrogen are protons, and 'deuterium' is known as the resulting nucleus.

Deuterium is, as discussed in Chapter 1, a hydrogen isotope whose nucleus is composed of a proton and a neutron and which makes up approximately 0.02% of natural hydrogen gas. The extra positive charge must go somewhere because the neutron holds no charge, and it is carried off by the emission of a positron (which is the same as an electron but with a positive charge) and a neutrino (a very small neutral particle). The ground state energy of the deuterium nucleus is slightly lower than that of the two protons, so we would have predicted that many years earlier, all the protons in the universe would have been fused into deuterium nuclei in the same way that nearly all hydrogen atoms form hydrogen molecules. The reason that this has not occurred is because of the electrostatic repulsion between the two protons that are positively charged.

One location where temperatures as high as a million degrees occur naturally is the sun, and the mechanism that keeps the sun shining is actually nuclear fusion. In addition to the two protons that form deuterium, several other fusion processes occur, and the endpoint is the most stable nucleus of all, that of iron. Fusion is also one of the concepts behind nuclear weapons, such as the 'hydrogen bomb.' In this case, the ignition from a nuclear explosion produced using atomic fission is achieved, which will be addressed shortly.

This heats the material to a temperature high enough to cause fusion, after which it is self-sustaining and results in a massive explosion. For over fifty years, nuclear engineers have been aiming to generate controlled fusion power that could be used for peaceful purposes.1 The technical challenges are enormous, and the machines needed to produce and sustain the necessary temperatures are enormous and amount to an investment of several billions of pounds.

International partnerships such as the JET2 project have been developed, and it is now assumed that within the first half of the twenty-first century, a computer capable of generating substantial quantities of fusion power would be installed. Towards the end of the twentieth century, rumors of 'cold fusion' created some interest, suggesting fusion energy was released without first providing heat. This work has been widely debunked, but some attempts in this direction continue.

Green Power

Over the last twenty or so years of the twentieth century, and since then, we have become increasingly aware of the fact that our exploitation of the Earth's energy supplies has given rise to major pollution-related problems and the like. Any of the initial concerns centered on nuclear technology, where the inevitable radiation that follows nuclear operations and the handling of radioactive waste products represent risks that some believed could not be handled. This was compounded by a small number of very large nuclear accidents, notably that in Chernobyl in Ukraine, which released a substantial amount of radioactive material across Europe and beyond.

However, more recently, the long-term implications of more conventional energy production approaches have become apparent. The threat of climate change correlated with 'global warming' is chief among these: there are clear signs that the combustion of fossil fuels contributes to a rapid increase in the temperature of the Planet, which in turn could contribute to the melting of the polar ice caps, a consequent rise in sea levels and the flooding of important parts of the populated areas of the Earth. There is also the possibility of a runaway mechanism in which, before the Planet was fully uninhabitable; heating would result in further heating.

There have been a rapid increase in interest in alternative,' renewable' forms of energy production in the face of such expected disasters. In this segment, we will first address how quantum physics plays a role through the 'green-house effect' in triggering global warming and how it can also lead to some of the sustainable alternatives.

The greenhouse effect is so-called because it mimics the processes found in many gardens that regulate a type of glass greenhouse activity. Sunlight passes without being absorbed through the translucent glass and hits the earth and other greenhouse contents, warming them up. By emitting heat radiation, the warmed objects then begin to cool off, but this has a much longer wavelength than that of light and does not move easily through the glass, which reflects most of the heat into the greenhouse.

This process continues until the glass has warmed up to the point that it radiates outward as much strength as the sunlight that comes through it. Convection supports the above process: air is heated at the bottom of the greenhouse, becomes less dense, and rises to the top of the greenhouse, where it helps to steam the glass as it cools and then falls back downwards.

CHAPTER 4: METALS AND INSULATORS

Those of us lucky enough to know how important electricity to modern life is to experience it. It gives us the power we use to illuminate our homes and our streets, cook our food, and drive the computers that process our data. The purpose of this chapter is to illustrate how all of this is a manifestation of quantum physics principles and, in particular, how quantum physics enables us to understand why the electrical properties of different solids can vary from metals that conduct electricity readily to insulators that do not.

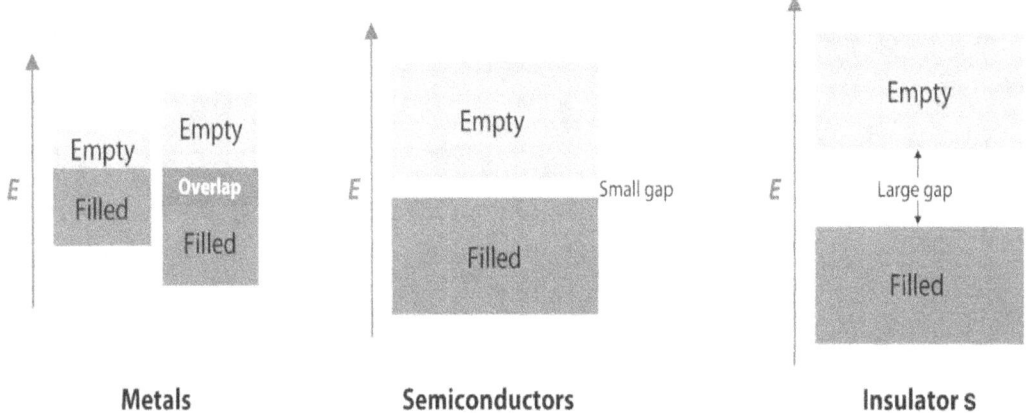

Metals   Semiconductors   Insulator s

Chapter 5 expands this topic to 'semiconductors'-materials with the necessary properties to allow us to develop the computer chips at the center of our information technology. First of all, I must stress that electricity is not in itself a source of energy, but rather a means of transferring energy from one location to another. In a power station, electricity is produced, which in turn gets its energy from some sort of fuel (e.g., oil, gas, or nuclear material) or from wind, waves, or sunlight. Quantum physics plays a part in some of these processes, too, as we saw in Chapter 3. Electricity comes to us in the form of an electric current that flows through a network of metal wires extending from the power station to the machine I use to write this portion through the plugin the wall.

An electric circuit consists of a battery that moves an electric current around a resistor-containing circuit. We need to have some knowledge of how this works and what these words mean. Next, the battery consists of several 'electrochemical cells' that use a chemical process on opposite ends of each cell to produce positive and negative electrical charges. This can then exert a force on any mobile charges connected to them, and the 'voltage' produced by the battery is called their capacity for doing so. First, the connecting wires are made of metal, and metals are materials that contain electrons that can travel freely inside the material (as I will explain in some detail shortly). When a wire is attached to a battery, electrons near the battery's negative terminal undergo a repulsive force that pushes them through the wire; they move around the circuit until they enter the positive terminal to which they are drawn; then, they travel through the battery and emerge at the negative terminal where the process is repeated.

As a consequence, a current flows through the circuit, and we should note that the traditional direction of current flow is opposite to that of the electrons since the electrons bear a negative charge. The reasoning for this is simply because the principle of electric current was developed, and before electrons were discovered, the traditional concept of positive and negative charge was established. The current passing through a resistor, as its name suggests, is a device that resists the current flow; its ability to do so is determined by a property known as its 'resistance.'

The voltage required to move a given current through a given resistor is proportional to the current and resistance size; this relationship is known as the 'law of Ohm,' which we will address in more detail towards the end of this chapter. All materials exhibit some resistance to electric current, apart from superconductors (to be discussed in Chapter 5), although the resistance of a standard copper wire is very weak. Resistors are often made from special metal alloys, built to have considerable resistance to current flow; some of their energy, which is transformed into heat, is lost by the current that flows through them. This is the mechanism that underlies the operation of any electric heater, such as that contained in or used to heat a room in a kettle or a washing machine.

What about the Ions?

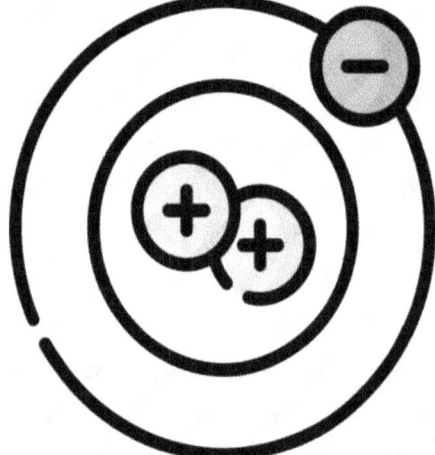

We have so far assumed that electrons are free to travel unhindered everywhere in the metal, but we realize that atoms are made of all solids. We might fairly assume that one or more electrons would be able to travel easily from one atom to the other in the highest energy shell, and we have ignored any interaction between them and the ions as a first approximation. However, because the latter carries a net positive charge, which should interact strongly with the negatively charged electrons, it is difficult to see how we can justify ignoring the ions entirely. Therefore, we might expect an electron seeking to pass through the metal to undergo a series of collisions with the ions, which would significantly hinder its motion, preventing the substantial flow of current.

Consider attempting to walk directly through a dense forest as an analogy: bumping into trees will continuously hamper your progress, slowing you down or even stopping your progress entirely. Why isn't anything similar happening with electrons? There are two explanations for this: the first is due to quantum mechanics and the fact that electrons have wave properties; the second is that their ions are arranged in a regular, periodic pattern since they are made up of crystals as a main characteristic of solids. In order to determine the electrical properties of solids, we can now see how these characteristics interact.

Although we may think of them as very exotic objects, such as expensive gemstones or crystals that have developed carefully in a science lesson in school, we are all familiar with crystals. It can come as a surprise to learn that many solids are crystalline, including metals. We will soon return to this stage, but first, we will look at some of the crystals' key properties and how they are expressed in their atomic structure. For their flat faces, sharp edges, and periodic forms, crystals are noted. In addition, these properties, particularly the normal form, are retained if a crystal is split into one or more pieces. This led to the discovery that crystal forms were a result of regularities in their atomic structure when the atomic composition of matter was discovered in the nineteenth century. In other words, a vast number of similar building blocks with atomic dimensions are made of crystals.

While many materials have a crystalline structure, this is not always immediately evident since a sample often does not consist of a single crystal but consists of a large number of crystalline 'grains' that are randomly oriented. These grains, usually one micrometer (106 m) in size, are small on a daily scale but are around a thousand times the size of a normal atom. We can presume that a solid consists of a single crystal and that it is rational to assume that current will flow from one grain to another in a typical wire if we can understand how a metal crystal can conduct electricity; this hypothesis is well confirmed experimentally.

A bit more about Metals

The fact that an electron's energy in a half-filled band is autonomous of the wavelength of the wave means that, without interfering with it, the wave function of an electron in the metal may have the form of a traveling wave that can pass through a crystal lattice. Metals would be ideal conductors of electricity if this were the whole story. However, although metals conduct electricity very well in practice, they still exhibit some resistance to current flow. The explanation for this is that the periodic atomic structure of a crystal is never entirely flawless, and the imperfections impede the flow of the current.

There are two major types of imperfection commonly encountered. Impurities are the first to be discussed, i.e., atoms of a different form from the main material component. These would usually be dispersed through the crystal more or less randomly, disrupting its periodicity at these points. The second imperfection happens because the ions are continually shifting because of the effects of temperature: some of them can be a large distance away from their normal positions at any moment so that the periodicity of the crystal is disrupted again. The net effect of all this is that, for the reasons discussed above, while electrons in metals move very freely through the crystal, they are distributed from time to time by impurities and thermal defects.

Usually, before encountering an impurity or thermally displaced ion, an electron travels a distance of several hundred ion spacings, but when it interacts with such a defect, it loses its forward momentum and moves in a random direction. The electric force acts on the electron at the same time to drive it forward again in the direction of the movement of the current. There is also tension between the electrical force driving forward the electrons and an efficient force associated with the scattering of the defect that attempts to resist this. As a consequence, in relation to the electric field and thus, the voltage applied, the size of the current flowing through a given sample increases. This effect is a familiar electrical conduction property known as 'Ohm's law.' The magnitude of the resistance to the current is also determined by the number and size of the defects that provide the dispersion.

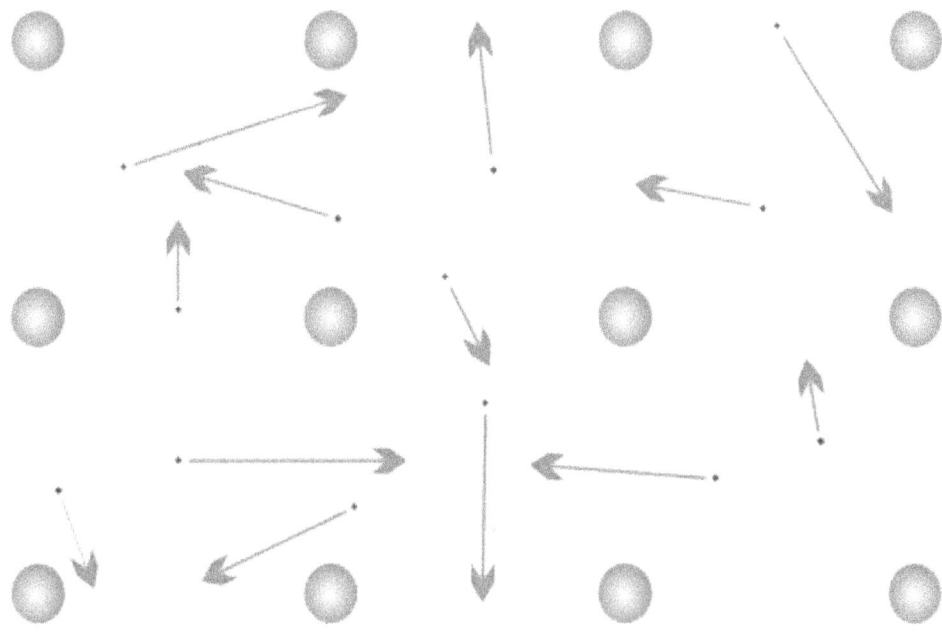

Those associated with thermal motion are typically the most important for relatively pure samples at room temperature. We want to engineer materials that resist current flow very strongly for some applications, such as electric heaters. This can be achieved by alloying two metals into a periodic crystal lattice of another, which can be considered adding a large proportion of impurity atoms of one kind.

This high impurity density then very strongly disperses the electrons and creates a completely independent resistance of temperature. In this chapter, we have seen how quantum physics is important for understanding the properties of metals and insulators and why they vary so significantly. In the next chapter, we will address semi-conductors to see how quantum physics plays a role in deciding semiconductors' properties, which are important for information technology that plays such a large role in modern life.

CHAPTER 5: SEMICONDUCTORS AND COMPUTER CHIPS

We saw in the last chapter how the dramatic contrast between metals and insulators was a function of the interaction between the electron-associated waves and the crystal's periodic array of atoms. As a consequence, the permitted energies of the electrons form a collection of bands separated by gaps. If there are enough electrons in the solid to only fill one or more bands, they do not respond to an electrical field, and the substance is typically an insulator. In comparison, in a typical metal, the highest occupied energy band is only half full, and the electrons react readily to an applied field, creating a current flow.

This chapter explains the properties of a class of materials that are known as 'semiconductors' that lie between metals and insulators. Semiconductors have an even number of electrons per atom, like insulators, and are thus only sufficient to fill many bands. The distinction is that the size of the distance between the top of the maximum complete band and the foot of the next empty band is fairly small in semiconductors, meaning not much greater than the energy at room temperature associated with an electron's thermal motion. There is then a large likelihood of thermal excitation of certain electrons from the complete band into the empty band. This has two electrical conduction effects.

Next, the electrons excited in the upper band (known as the 'conduction band') will pass freely through the metal, bringing an electrical current, as there are plenty of empty states available to move in.

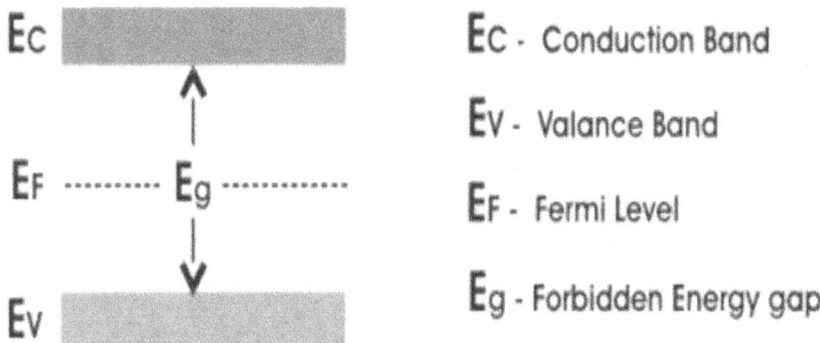

E_C - Conduction Band

E_V - Valance Band

E_F - Fermi Level

E_g - Forbidden Energy gap

Second, the empty states left behind in the lower band (known as the 'valence band') are accessible in this band to the electrons, so they can pass freely and bring current as well. Thus, both bands contribute to the current flow, and a perfect insulator is no longer the material. We are now considering in a little more detail the behavior of the nearly full lower band. We will discover that its properties are not negative electrons but only charged particles. First, to see how this works, we remember that an equivalent number of electrons pass in opposite directions in a complete band.

We see that an imbalance occurs if one of these electrons is removed, and the net effect is a current similar to but opposite to that associated with the missing electron, but this is just the current that would result from a single positive charge traveling at the same velocity as the missing electron. The effect of applying an electric field to the device exerts the same force on all electrons, causing their velocities to change by the same amount, so the net change is again equal and opposite to what the missing electron would have experienced. The behavior of a collection of electrons with one removed is, therefore, the same as that expected from a particle that has all the properties of a single electron, except that its electrical charge is positive.

The p–n Junction

By joining a piece of p-type semiconductor to one of n-type to form a 'p-n junction,' one of the simplest devices to exploit the above characteristics is made. Such a junction is found to operate as a current rectifier, which means that if the current flows in one direction (from n to p), it is a strong electrical conductor but provides a large resistance to current flow in the opposite direction. We first consider the junction area where the n-type and p-type meet to understand how this comes about. An electron entering this region will make a transition in the lower band into a vacant level, thus removing the electron along with a hole; as discussed above, we say that the electron and hole have been annihilated.

As a consequence, in the junction area, there is a deficiency of both electrons and holes, and the charges on the ions are not fully canceled, leaving narrow bands of positive and negative charges on the junction's n-type and p-type sides, respectively; these charges are referred to as 'space charges.' Consider applying a voltage that would push a current from p to n: the holes move in the direction of that current in the p-type material, while the electrons move in the reverse direction in the n-type.

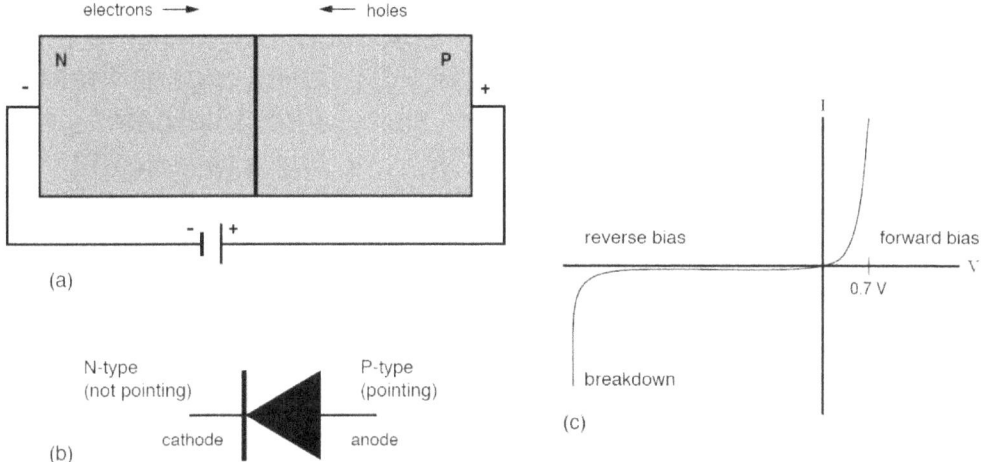

(a)

N-type
(not pointing)

P-type
(pointing)

cathode

anode

(b)

reverse bias

forward bias

0.7 V

breakdown

(c)

This leads to a rise in the number of charging carriers in the junction area as well as a consequent decrease in the size of the space charges. They annihilate some of the holes as the electrons reach the middle of the junction onto the p-type material, and as holes pass through the n-type material, electrons are equally annihilated. They are replaced by those drawn in from the external circuit, and a current flows as the electrons and holes disappear from the n-type and p-type material, respectively. The junction is defined as being 'forward-biased' when a voltage is applied to drive a current in this direction. On the other hand, if a voltage is applied in a direction that appears to push a current from N to P, the electrons and holes are drawn away from the central region; as a result, the space charges are increased, and it becomes more difficult to cross the junction for charge carriers.

The current will then not continue to flow, and the junction is said to be 'reverse-biased' in this configuration. Thus, when a voltage is applied in one direction but not when it is applied in the other, a current flows, which means that a p-n junction has only the above-mentioned rectifying properties. In the domestic and industrial applications of electricity, rectifiers like those made from p-n junctions have many applications. In a power station, electricity is generated by a generator powered by a spinning motor. One effect of this is that the electricity produced is 'AC,' which means that any time the motor rotates, the voltage generated alternates from positive to negative and back, usually fifty times a second. For certain applications, such alternating currents are perfectly satisfactory, such as room heaters and electrical engines used in washing machines, etc.

However, some systems need a power source where the current always flows in the same direction; this is important, for example, when we charge a battery in a car or use a charger supplied with a cell phone. Rectifiers can be used to convert AC to DC (direct current) based on p-n junctions, where we see that current flows during only half of the cycle when an alternating voltage is applied to a p-n junction, so the resulting output would have only one sign, while half of the AC cycle is zero. If we use four connected rectifiers at all times, the output voltage again has the same sign but is now on during the loop. In comparison to the 'half-wave rectification' accomplished by a single rectifier, this is known as 'full-wave rectification.'

Rectifiers are part of all battery chargers based on p–n junction units. A steady DC source is often required: this can be supplied by a charged battery or can be generated by smoothing the output using an electrical part known as a 'capacitor' from a rectifier.

The Transistor

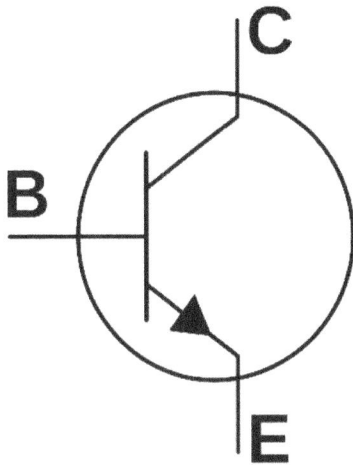

Essentially, the semiconducting properties of silicon are the foundation of all modern information and communication technologies. Its use as 'a a transistor' is especially important. A transistor is a device that can be used to convert a small signal into one of a similar shape but strong enough to control a loudspeaker or similar device, such as that detected by a radio receiver. As a controlled switch, a transistor may also be used and, as such, plays an important role in the operation of computers and other digital devices. We shall define the construction of a transistor in this section and illustrate how it can be used in each of these ways.

We apply a positive voltage between the emitter and collector to operate a transistor. We see that holes can easily flow across the emitter-base junction from the earlier discussion of p-n junctions because it is forward biased, but we expect no current to flow between the base and the collector because it is reverse biased. A significant aspect of a transistor design, however, is that the base region is purposely made very thin and is lightly doped with impurities so that at least some of the holes can pass through the base from the emitter to the collector without meeting any electrons to be recombined with.

A current will flow around the circuit as a result. Now, consider the result of injecting electrons into the base, which corresponds to drawing a current from it. Some of the electrons will cross the base and pass between the base and the emitter through the forward-biased junction, while others will be annihilated by combining holes that flow from the emitter to the collector through the base.

The resultant current going through the collector would be very much greater (typically one hundred times) than the base current if the system is properly constructed. Tiny changes in the base current cause correspondingly large changes in the collector current, so we have a 'current gain.' This gain depends on the densities of holes in the p-type regions and of electrons in the base region and the dimensions of the base; if the voltage of the base-emitter is not too large, the gain for a given transistor is constant. The collector current reaches a 'saturation' value as greater voltages are added to the base and stay at this level even if the base voltage is further elevated.

By constructing the circuit, the current gain can be converted to a voltage gain. An input voltage drives a current into the base of a transistor through a resistor and then enables the supply voltage (i.e., a battery or other power source) to push a current from the emitter to the collector through the transistor and then through a second resistor. Therefore, through it, an output voltage equal to the current through this resistor occurs.

As a consequence, the output voltage is equal to the voltage of the input, and we have a gain in voltage. Now we are turning to the topic of how to use a transistor as a controlled switch. The theory is that we add either a zero voltage or a significant voltage to the input resulting in either a very small current or a relatively large (i.e., saturation) current flowing through the emitter. Depending on the base voltage's size, the emitter current, and hence the output voltage, is thus turned on or off.

For the design of basic operating units in a digital machine, this theory can be extended. First, we notice that a series of 'binary bits' can represent any number, each of which can have a value of 1 or 0.

'10' represents 1 2 + 0 1 (i.e. 2) following this convention, while '100' represents 1 2 2 + 0 2 + 0 1 (4) and so on. Therefore, for instance, 101101 = 1 32 + 0 16 + 1 8 + 1 4 + 0 2 + 1 1 = 45

A binary bit can be used to represent any physical system that can occur in any of the two states. In applying this concept in the sense of transistors, we embrace the convention that a voltage greater than a certain threshold represents 1, whereas a voltage greater than a certain threshold represents 0. The numerical value of the binary bit expressed by the output voltage in Figure 5.7(b) will therefore be 0 if the input voltage is 0 and 1 if the output voltage is 1.

Let us consider one of the basic computer operations as a further example: the 'AND' gate, which is a system in which an output bit equals one if and only if there are two input bits each.

Such a unit can be built using two transistors. Only if both base currents are high enough will a current flow through both transistors. Both input voltages must be high for this to happen, meaning that the digits represented by them are both 1. The digit defined by the output voltage, in this case, will also be 1, but otherwise, it will be 0. This is just the property that an AND gate needs. The other simple operations such as 'OR' can be configured to perform similar circuits, where the output is one supplied, either input is 1, and 0 otherwise.

All computer operations are made up of combinations of these and other related basic components, including those used in arithmetic. Computers designed shortly after the transistor was created in the 1950s and 1960s were actually constructed along the lines mentioned above from individual transistors.

However, large numbers of transistors were required to meet their requirements as their operation became more sophisticated. The invention of the 'integrated circuit' was a significant development in the mid-1960s, in which many circuit components such as transistors and resistors, as well as the equivalents of the wires connecting them, were contained on a single semiconductor piece, known as a 'silicon chip.' It became possible to reduce the size of the individual components as the technology advanced and thus have more of them on each chip. This had the added benefit that it was also possible to reduce the switching times, so machines have become increasingly more efficient and faster over the years.

The Pentium 4 processor in the machine on which I write this text consists of a silicon chip in the region of about one square centimeter; it has around 7.5 million circuit components, many smaller than 10^{-7} m in size, and the basic switching time or clock speed is about 3 GHz (i.e., 3×10^9 operations per second). However, 10^{-7} m is several hundred times the atomic separation, so it is still possible to view each part as a crystal, and the transistors in a silicon chip work according to the same principles of quantum physics as we have mentioned in this chapter.

The Photovoltaic Cell

A photovoltaic cell is a semiconductor-based system that converts the energy from sunlight into electricity. Because all the energy comes from the sunlight that will, in any event, hit the Earth, it does not contribute to the greenhouse effect and does not absorb any of the fossil or nuclear fuel reserves on Earth. Over the years, numerous such devices have been developed, and the research is motivated by the desire to improve this non-polluting type of energy to the point that a significant portion of human energy consumption can be fulfilled. The semiconductors are all made of photovoltaic cells.

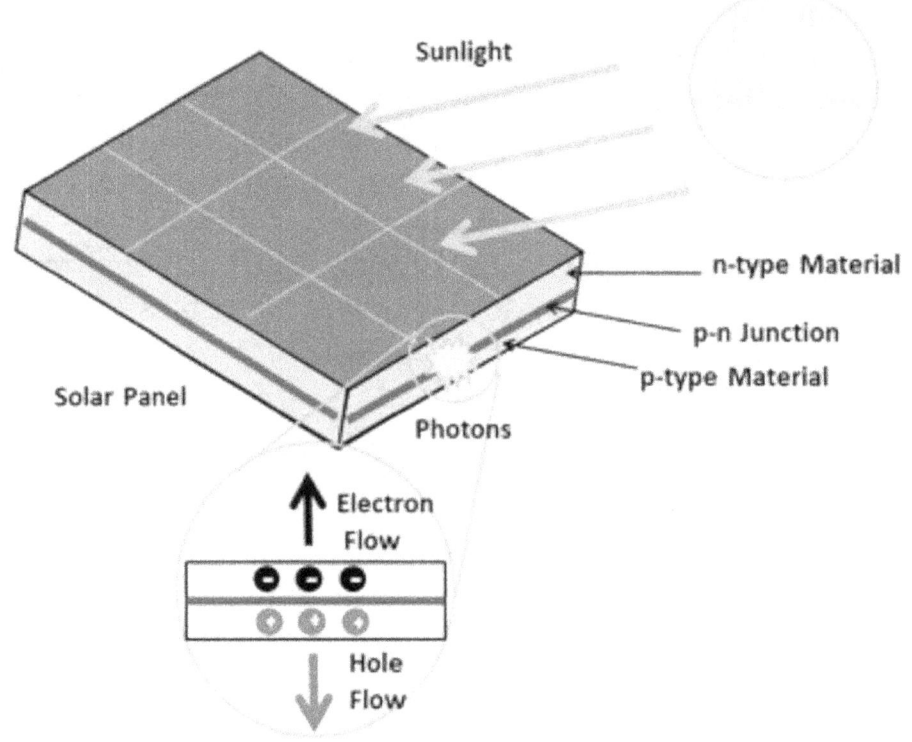

Sunlight

n-type Material

p-n Junction

p-type Material

Solar Panel

Photons

Electron Flow

Hole Flow

If a photon of the correct energy reaches a semiconductor, it may cause the upper band to be excited by an electron, leaving a positive hole in the lower band. We need the electron and the hole to shift away from each other and drive a current into an external circuit in order to generate a voltage. Using a p-n junction is one method of doing this. As mentioned above, there are very few charge carriers in the junction region where the p-type and n-type materials meet since the electrons and holes cancel each other out; and on the n-type and p-type sides of the interface, respectively, there are excess positive and negative charges.

The photons can be absorbed if we shine a light on this junction field, exciting electrons from the valence to the conduction band and producing an electron-hole pair. These can easily recombine, but there is a large likelihood that the electrons will instead be accelerated by the electrostatic forces acting on them into the n-type region, while the holes are similarly accelerated into the p-type material. Consequently, to charge a battery, the unit will push a current through an external circuit.

CHAPTER 6: SUPERCONDUCTIVITY

We saw in Chapter 4 how the fact that electrons can behave as waves makes it possible for them to travel through a perfect crystal without bumping on the way through the atoms. Provided empty states are open, as in a normal metal, the electrons will react to an applied field, and a current will flow, whereas the presence of an energy gap in an insulator implies that there are no such empty states and thus no flow of current.

In practice, we also saw that the current flow through a metal encounters some resistance because all actual crystals involve imperfections related to the thermal displacement of atoms from their regular crystal positions and the substitution by impurities of some of the atoms. We will address another class of substances in this chapter, known as 'superconductors,' in which resistance to current flow fully disappears, and electric currents will flow once they have begun. Ironically, we can find that this behavior often results from the existence of an energy gap that has certain parallels to the energy gap that prevents current flow in an insulator, as well as major variations.

A Dutch physicist, Kamerlingh Onnes, discovered superconductivity, more or less by mistake, in 1911. He was performing a program of calculation of the electrical resistance of metals at temperatures reaching absolute zero, which had recently become accessible thanks to technical advances in the liquefaction of gases. In fact, helium liquefies at normal pressure at just a few degrees above absolute zero and can be cooled much more by reducing its pressure using vacuum pumps. Onnes finds that as the temperature is decreased, the resistance of all metals decreases, although, in most situations, there is still some resistance to current flow at the lowest possible temperatures and, by inference, at absolute zero. In terms of the model that we built in Chapter 4, we can understand this behavior.

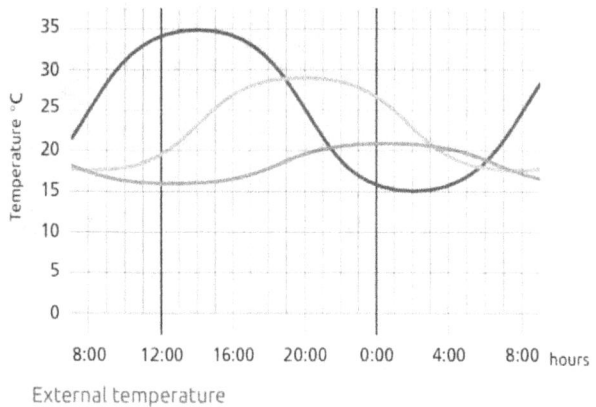

External temperature

Environmental side of roof with mineral fiber insulation

Environmental side of roof with wood fiber insulation

The thermal displacement of the atoms from their average locations as the temperature decreases, and there is less possibility of these interfering with the electrons; however, the resistance due to impurities is temperature-free and is still present at absolute zero. It was when Onnes concentrated his attention on the lead of the metal that unexpected results arose. This does not suggest that it actually becomes much smaller than it is zero for copper. In research, this is quite rare. Normally, "zero" means much less than any comparable quantity in the same way as "infinite" means something much greater than other comparable quantities, but "zero resistance" really means what the word implies in this case. This is another result, as we can see, of quantum mechanics influencing the everyday world.

'High-Temperature' Superconductivity

We stated above that when Kamerlingh Onnes used liquid helium to cool the material down to below 4 K; superconductivity was first observed in the lead. Superconductivity was observed in some metals and alloys in the seventy-five years that followed this discovery, but the highest critical temperature was less than 23 K, and the use of liquid helium was still needed to achieve this temperature. Helium is, still now, a gas that is difficult and costly to liquefy. It must be surrounded by two vacuum flasks with the space between them filled with liquid nitrogen to keep the liquid from boiling. Thus, superconductivity was considered a pure-science subject with only the most specialist applications until the 1980s.

Later, in 1986, superconductivity for 'high-temperature' came along. J. Two scientists who worked for IBM in Zurich were Georg Bednorz and Karl Alex Müller. They realized the promise of a superconductor that would work at temperatures higher than that of liquid helium and started a program of testing various materials almost as a spare-time task to see if any would live up to this dream.

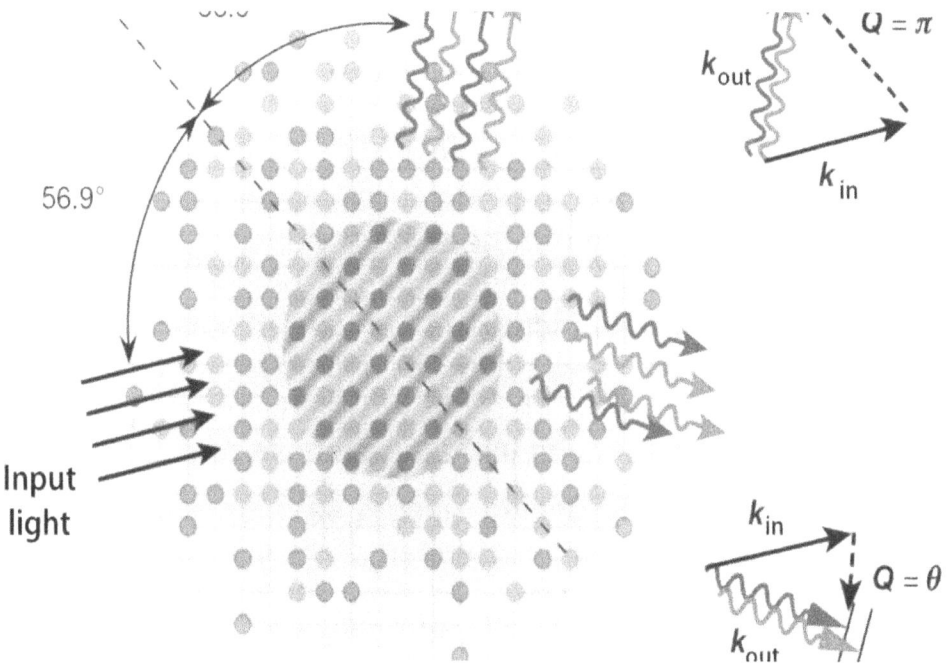

When they turned their attention to a particular compound of lanthanum, bismuth, copper, and oxygen, they were probably as shocked as anyone else, and they discovered that its electrical conductivity fell dramatically to zero when it was cooled below 35 K-which, while still a very low temperature, is more than one and a half times the previous record. In January 1987, a research team at the University of Alabama, Huntsville, substituted yttrium for lanthanum in a compound close to that found by Bednorz and Müller and discovered that the compound superconducts up to 92 K.

Not only was this another major development on the temperature scale, but the nitrogen boiling point, which is 77 K, has also passed a significant milestone. This meant that it was now possible to demonstrate superconductivity without the use of liquid helium. It is much simpler to manufacture liquid nitrogen than liquid helium, more than ten times cheaper and can be stored and used in a simple vacuum flask. Superconductivity could be examined for the first time without costly, specialist equipment; on the laboratory bench, superconducting phenomena such as magnetic levitation that had previously been seen only through many layers of glass, liquid nitrogen, and liquid helium could be seen.

Progress has been less dramatic since 1987. For a composite of the elements mercury, thallium, barium, calcium, copper, and oxygen, the highest known transition to the superconducting state occurs at normal pressures at 138 K; under intense pressure, its transition temperature can be further increased to over 160 K at a pressure of 300,000 atmospheres. They have been named 'high-temperature superconductors' due to the fact that the transition temperatures of these compounds are so much higher than those previously found.

This title is potentially misleading since it seems to suggest that at room temperature or even higher, superconductivity can exist, which is definitely not the case. The maximum superconducting temperature, however, was increased between 1986 and 1987 from 23 K to 92 K, i.e., four times; if another factor of three could be achieved, the dream of a superconductor at room temperature would have been achieved. We may have predicted that the advance to liquid nitrogen temperatures would have significantly improved the potential for practical superconductivity applications, but these were less drastic than originally planned.

For this, there are two main explanations. Next, what is known as 'ceramics' are the materials that constitute high-temperature superconductors. This ensures that they are physically identical to other ceramics (such as those used in kitchens) because they are hard and brittle, making them very hard to produce in a shape that is ideal for replacing metal wires.

The second issue is that the overall current that can be sustained by a high-temperature superconductor is rather too small to be practical for electricity transport or the development of strong magnetic fields. This is still, however, a field of active research and development. For example, in the early years of the 21st century, the design of motors based on high-temperature superconductors entered the prototype stage. Their greatest ability is where high power is required, combined with low weight: an electric motor to power a boat, for example.

Flux Quantization and the Josephson Effect

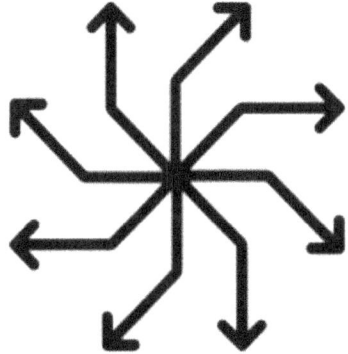

We have shown that there are Cooper pairs in superconductors in which the electrons are bound together. As a consequence, it is possible to conveniently characterize the quantum mechanics of superconductors as the motion of such pairs rather than the individual electrons. In fact, such a pair can be thought of as a particle with a mass equal to twice the mass of the electron and a charge equal to twice the charge of the electron, traveling at speed equal to the pair's net velocity. From the pair's velocity and mass, the wavelength of the matter-wave associated with such a particle can be determined using the de Broglie relation.

This places constraints on the value possessed by the magnetic field through the loop for very subtle reasons: its flux always equals a whole number of times the 'flux quantum,' which is defined as Planck's constant divided on a Cooper pair by the charge. This works out as equal to a field of magnitude around two-millionths of the Earth's magnetic field passing across an area of one square centimeter.

CHAPTER 7: Spin Doctoring

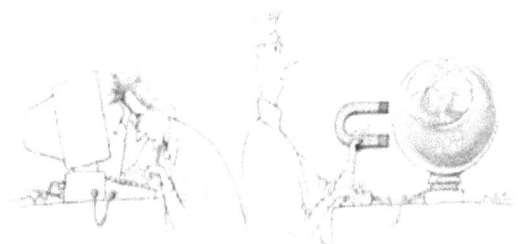

There has been a growing interest in the application of quantum physics to the processing of knowledge in computers, for example, over the last decade of the twentieth century and ever since. Chapter 5 reveals that modern computers are built on semiconductors, which are, in turn, governed by the laws of quantum physics. Despite this, these computers are still widely referred to as 'classical' since the calculations are done in a completely classical way, whereas quantum physics underlies their operation. In order to better understand this, we must first note that all data on a standard machine is represented by a set of binary 'bits' that can equal either 1 or 0.

How these are interpreted is unrelated to the manner in which equations are manipulated to perform. However, quantum mechanics is central to the actual operations of computation in quantum information processing: information is expressed by quantum objects known as 'qubits' where behavior is governed by quantum rules. A qubit is a quantum system that can be in one of two states (like a classical bit) and can represent 1 and 0, but a qubit can also be in what is referred to as a 'quantum superposition' of these states, in which both 1 and 0 are concurrently in some way.

What this implies should soon become clearer as we consider some particular instances where we can see that some items that are classically impossible can be achieved by the quantum processing of information. While several different quantum systems could be used as qubits, we will restrict our discussion to the electron spin example. We found that electrons, and indeed other fundamental particles, have a quantum property that we referred to as 'spin' in earlier chapters.

By this, we say that a particle behaves as though it were rotating around an axis in a manner reminiscent of the Earth's rotation or that of a spinning top. Like too often occurs in quantum mechanics, if we try to take it too literally, this classical model is best thought of as an example, and difficulties arise. For our purposes, the main thing to remember is that spin determines a direction in space, which is the axis around which the particle 'spins,' and that when we calculate the spin of a fundamental particle, such as an electron, we find that it often has the same magnitude, while its direction is either parallel or anti-parallel to the rotation axis.

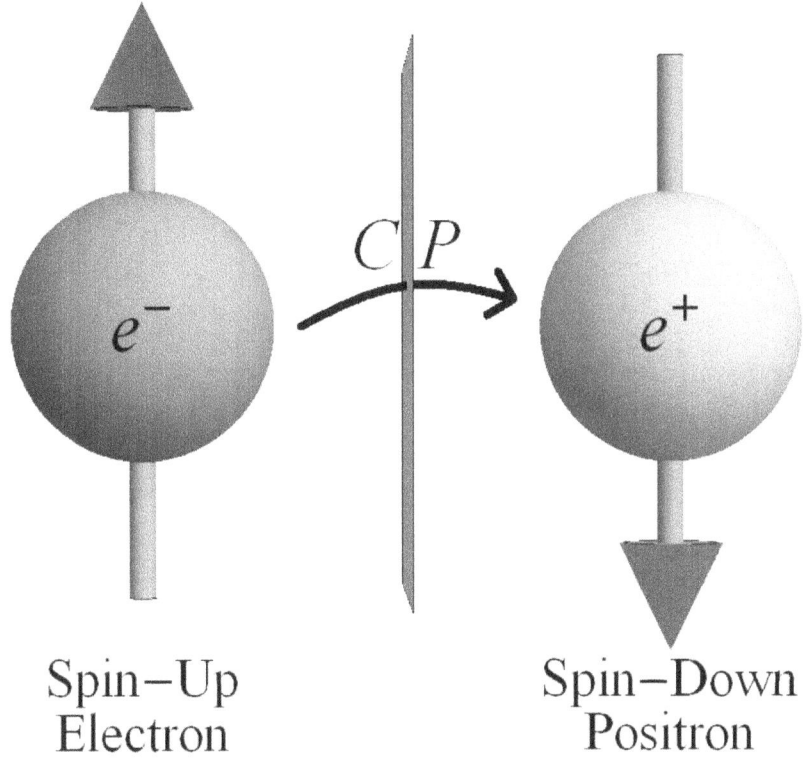

$C\ P$

Spin–Up
Electron

Spin–Down
Positron

As a shorthand, we can assume that either 'up' or 'down' the spin is pointing;1 and we saw in Chapter 2 that these two possibilities play an important role in deciding the number of particles permitted to inhabit any given energy state by the exclusion principle. Therefore, we see that spin has at least one of the necessary properties of a qubit: it can exist in one of two states that can be used to describe the binary digits 1 and 0. Now we are going to try to explain how it can also be placed in a state of superposition and what this means. What do we, we may wonder, mean by 'up' and 'down'? The electron can certainly not be influenced by such a concept, which relies on our experience of living on the surface of the Earth and, in any case, the directions that we think of as 'up' and 'down' change as the Earth rotates.

Why shouldn't we be able to calculate spin relative to a horizontal axis, for example, so that it is either 'left' or 'right'? The answer to this question is that we can measure spin relative to any direction we like, but we always find that the spin is either parallel or antiparallel to it once we choose such a direction. The act of doing such a measurement, however, destroys any data we may have previously had about its spin relative to some other direction. That is, the measurement appears to force the particle to reorient its spin so that either parallel or anti-parallel to the new axis is oriented.

How do we, in fact, calculate spin? The most straightforward approach is to use the assumption that there is also a related magnetic moment with every particle that possesses spin. By this, we say that an electron-like fundamental particle acts like a tiny magnet pointing along the spin axis. Thus, if we can calculate this magnetic moment's direction, the effect also informs us of the direction of the spin. One way to calculate this magnetic moment is to position the particle in a laboratory-generated magnetic field; if this field is greater when we move in, say, an upward direction, then a magnet pointing in that direction will move upward, while one pointing down will move downward.

Moreover, the size of the force causing this motion is proportional to the magnitude of the magnetic moment and hence of the spin, which can therefore be deduced from the amount the particle is deflected. Otto Stern and Walther Gerlach, two physicists, based in Frankfurt, Germany, first performed this technique in 1922. Via a specially built magnet that divided the particles into two beams, one corresponding to spin up and one to spin down, they passed a beam of particles 2.

Quantum Cryptography

Cryptography is the science of coding messages using a key or cipher so that they can be transmitted to another ('the recipient,' named 'Bob') from one person ('the sender,' historically called 'Alice') while remaining incomprehensible to an 'eavesdropper' ('Eve').

There are several ways to do this, but we will focus on one or two basic examples that explain the values involved and the contribution that can be provided by quantum physics. Suppose the message is the word 'QUANTUM' that we want to give. A simple code is simply to substitute each letter in the alphabet with the one that follows it unless it is Z that is 'wrapped around' to become A.

More generally, by substituting each letter with n letters later in the alphabet and wrapping around the last n letters of the alphabet to be replaced by the first n, we can encode any message. We would, therefore, have

Plain Q U A N T U M message
Coded using n = 1 R V B O U V N
Coded using n = 7 X B H U A B T
Coded using n = 15 F J P C I J B

This code is really easy to crack, of course. There are only twenty-six different possible values of n, and it would take only a few minutes to try them all with a pencil and paper; in a tiny fraction of a second, a machine could do this. The correct value of n will be defined as the only one that produces a sensible message; if the original message is relatively long, the chances of there being more than one of these are very small.

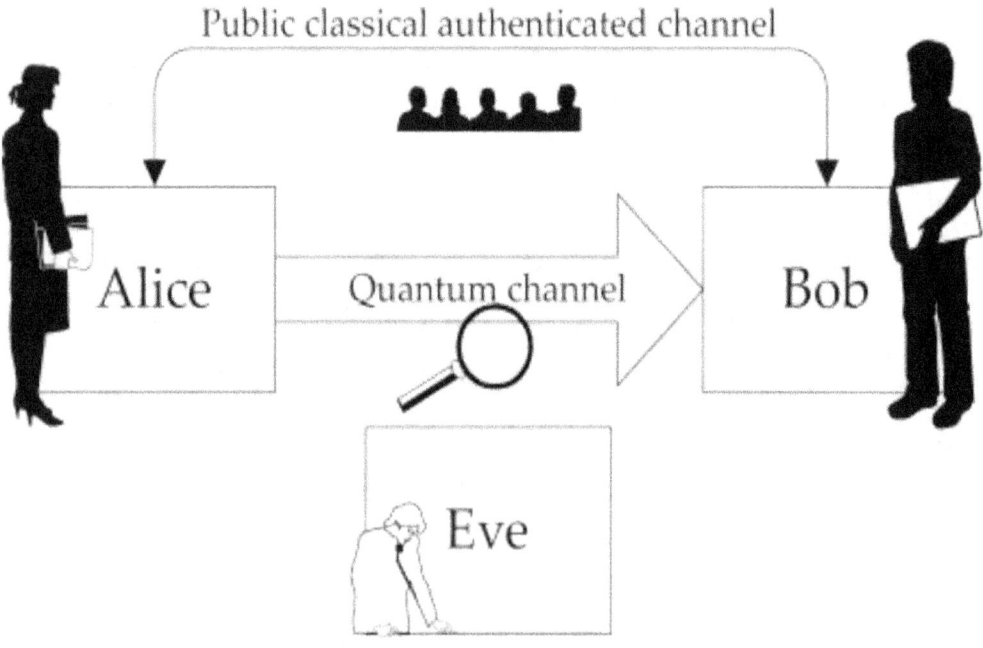

97

The use of arithmetic relies on a basic yet slightly more sophisticated method. First, we substitute a number for every letter in the post so that A becomes 01, B becomes 02, and so on, so that Z is defined by 26. We then add a known 'code number' to the message, which can be created as many times as possible to generate a number as long as the message by repeating a shorter number (known as the 'key' to the code). Underneath the note, this number is written, and the two rows of digits are applied to create the coded message.

In the instance below, where we choose the key to be 537, this procedure is illustrated.

Plain Q U A N T U M message

Numbers 1 7 2 1 0 1 1 4 2 0 2 1 1 3 3

Code number 5 3 7 5 3 7 5 3 7 5 3 7 5 3

7 0 9 6 3 8 6 7 9 5 5 8 6 6 6 Coded message

Alice gives Bob the last line, and he can retrieve the message by regenerating the code number and subtracting it from the coded message, given he knows the method and the values of the three digits. If Eve intercepts the message and attempts to decipher it, once she sees a meaningful message, she will have to try all the one thousand possible values of the key. A machine can still do this quite easily, of course. Such instances have a major characteristic in common, which is that the code key is much shorter than the message itself. Mathematical methods that can be used are far more complicated.

A message can be encoded in such a way that a present-day classical machine will have to operate for several years to be sure of cracking the code, using a key that consists of around forty decimal digits. Therefore, if in full secrecy, Alice and Bob could exchange a short message, they could use this to determine which key to use before sending the message, and then freely exchange coded messages, sure that Eve won't understand them. This does depend, however, on Alice and Bob knowing the key and Eve not having access to it. It is this protected key exchange that, as we shall now see, is enabled by the use of quantum techniques.

Quantum Computers

 The quantum computer is another instance of quantum information processing. In the present tense, we should note that it is incorrect to speak about quantum computers because the only machines that have been designed to date are capable of only the most trivial calculations that can be carried out more easily on a pocket calculator or even by mental arithmetic. Nonetheless, if the technical challenges could be solved, quantum computers would have the capacity to conduct certain calculations much faster than any traditional machine imaginable.

For this reason, in recent years, the promise of quantum computing has become something of a holy grail, and a great deal of scientific and industrial investment is being devoted to its growth. It remains to be seen if this will pay off or not. So how is it possible to manipulate the principles of quantum physics to this end, except in principle? A comprehensive discussion of this is well beyond the boundaries of this book, but we may expect to grasp some of the fundamental concepts involved. The first important argument is that a binary bit is not represented in a quantum computer by an electric current flowing through a transistor but by a single quantum entity such as a spinning particle. In the previous section, we saw an example of this when we discussed quantum cryptography.

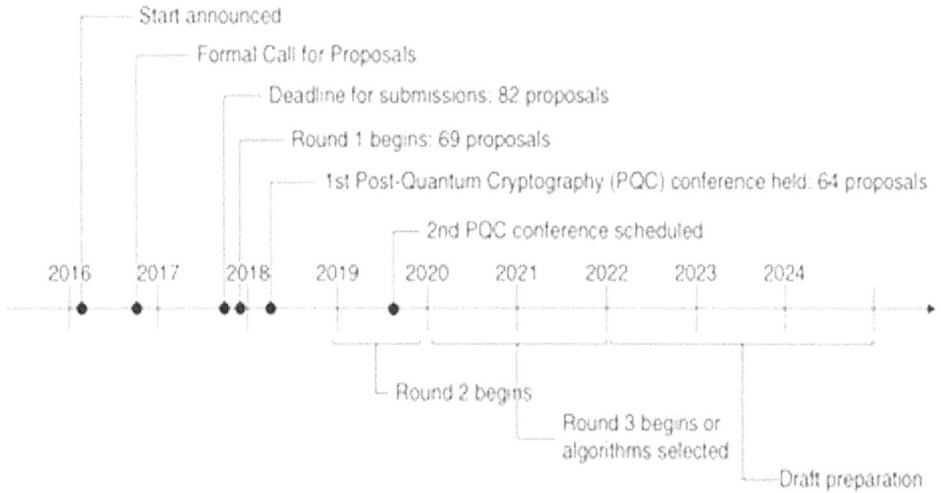

As before, we will assume that a particle with a positive spin in the vertical direction (spin-up) represents 0, while a negative spin component represents 1 (spin down). It is generally referred to as a 'qubit' when a quantum entity is used to represent a binary bit in this way. We consider how we can perform the 'NOT' operation as a first example, which is one of the simple Boolean operations that make up the computation and consists of replacing 1 with 0 and 0 with 1. Note that a particle that spins behaves like a small magnet.

This implies that it would want to turn like a compass needle to match up with the direction of the field if it is put in a magnetic field. The inertia of the spin would resist this motion, three but by applying a carefully regulated magnetic field to a spinning particle, the spin can be rotated at any known angle. If this angle is 180 °, for example, an up spin will be rotated to point down, and a down spin will be rotated to the up position, which is just what we do NOT need to reflect the process.

It can also be seen that by subjecting spinning particles to properly engineered magnetic fields, all the operations that a traditional computer performs on bits can be performed on qubits. Some of these include interactions between qubits, which is one of the obstacles to a quantum computer's practical realization.

What does it all Mean?

So what?

With wave-particle duality, we started our discussion of quantum physics. Light historically thought of as a type of wave motion often acts as if it were a stream of particles, while it was found that artifacts, such as electrons, which were once thought of as particles, had wave properties. We avoided any thorough explanation of these principles in the earlier chapters and instead focused on explaining how they are applied to model the action of atoms, nuclei, solids, etc.

We shall return to questions of theory and the philosophical problems of the subject in this chapter. A word of warning: this is a field of significant controversy, where many alternative approaches exist, which means that our debate is more about philosophy than physics. By considering the 'Copenhagen interpretation,' which is the traditional view among physicists, we will begin our discussion. Towards the end of the section, some alternative methods are briefly discussed.

A mineral calcite crystal is a less familiar type of polarizer: it is split into two beams as unpolarized light passes through this system, one of which is polarized parallel to a particular direction specified by the crystal, while the other is perpendicular to it. In one or other of these lights, unlike Polaroid, where half the light is lost, all the light emerges. It is important to remember that a calcite crystal is not like a filter that only makes a limited amount of light that has already been polarized in the right direction. Instead, with perpendicular polarization, it splits or 'resolves' the light into two components, and the sum of their strength is equal to that of the incident beam; regardless of its initial polarization, no light is lost. We will describe a polarizer such as a calcite crystal as a box where a beam of light enters from one side and emerges as two beams of perpendicular polarization from the other—the specifics of how all these works are not important to our intent.

Polarization is an electromagnetic wave property, but does it have any relation to light's particle model? By passing very weak light through a polarizer set up, we might test this: we will find photons (the light particles first described in Chapter 2) arising at random through the two output channels, corresponding to horizontal polarization (H) and vertical polarization (V). To confirm that the photons really can be regarded as having the property of polarization, we could pass each beam separately through other polarizers also oriented to calculate HV polarization.

We can find that all photons emanating from the first polarizer's H channel would emerge from the second one's H channel, and similarly, for V. This gives us an operational concept of photon polarization: we may assume that horizontally and vertically polarized photons are those that emerge from a polarizer's H and V channels, respectively, whatever this property might be. Thus, the properties of polarized photons are identical to those of spinning electrons described in Chapter 7 in certain respects, and another example of a qubit is a polarized photon.

The Measurement Problem

The above may be difficult to embrace, but it works, and if we apply the rules and use the map book properly, we can correctly determine predictable outcomes of measurement: the energy levels of the hydrogen atom, the electrical properties of a semiconductor, the product of a calculation carried out by a quantum computer and so on. However, this implies that we understand what 'measurement' means, and this turns out to be the most challenging and contentious topic in quantum physics interpretation.

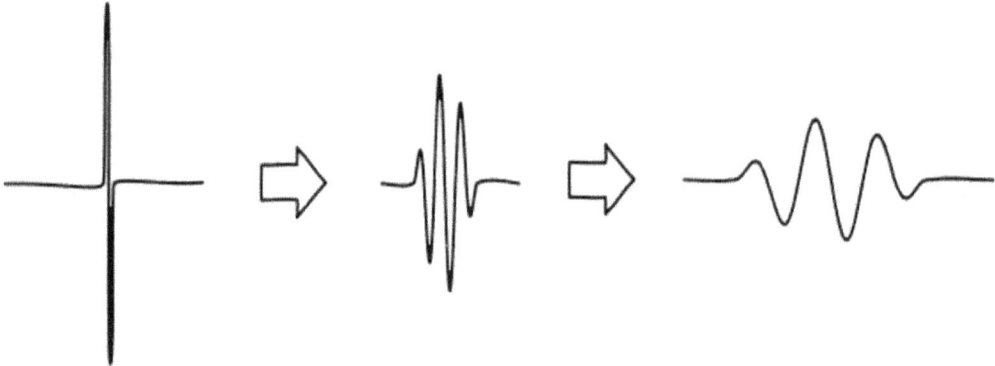

This is consistent with the positivist approach discussed earlier since we do not know that the photon has polarization in the absence of detection, so we can not conclude that it does. Therefore, by the presence or absence of a detector in the experimental arrangement, we seem to be able to divide the quantum universe from the classical world.

Alternative Interpretations

Subjectivism

To escape into 'subjective idealism' is one response to the issue of quantum measurement. In doing this, we simply agree that quantum physics means that an objective account of physical reality can not be given. Our personal subjective experience is the only thing we know that must be real: the counter may fire and not fire, the cat may be both alive and dead, but I definitely know what has happened when the knowledge enters my mind through my brain

.

For photons, counters, and cats, quantum physics may apply, but it does not apply to you or me! I don't know, of course, that the states of your mind are real either, so I'm in danger of relapsing into 'solipsism' in which only I and my mind have any truth.

Philosophers have long proposed that they could prove the existence of an actual physical universe, but science's aim is not to address this question but to provide a clear account of the current objective world. If quantum physics were to eventually ruin this mission, it would be ironic. Most of us would much prefer to explore an alternative path forward.

Hidden Variables

An understanding that rejects Bohr's positivism in favor of realism is based on what is known as 'hidden variables' (or 'naïve realism' as some of its critics prefer), implying that a quantum object has properties, even though they can not be observed. After Louis de Broglie, the first person to postulate matter waves, and David Bohm, who developed and extended these ideas in the 1950s and 1960s, the leading theory of this kind is known as the 'de-Broglie-Bohm model' (DBB). Both the particle position and the wave are believed to be actual properties of a particle at all times in DBB theory.

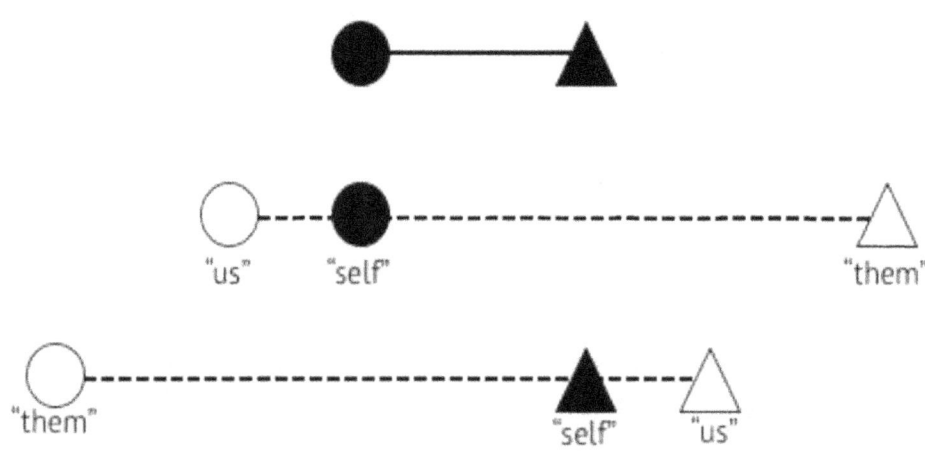

The wave forms according to the laws of quantum mechanics, and both the wave and the classical forces acting on it direct the particles. The direction taken by any particular particle is then fully decided and at this stage there is no ambiguity. However, various particles arrive at different locations depending on where they originate from, and the theory ensures that the numbers arriving at different points are compatible with the probabilities predicted by quantum physics. Consider the two-slit experiment as an example:

the form of the wave is determined by the shape, size, and location of the slits, according to DBB theory, and the particles are directed by the wave so that most of them end up in positions where the pattern of interference has high intensity, while none reaches the points where the wave is zero. As we have mentioned before, in a classical sense, the appearance of seemingly random, statistical effects from the action of deterministic systems is very familiar.

If we toss a large number of coins, for example, we can find that approximately half of them come down heads while the rest display tails, even though the action of any individual coin is controlled by the forces acting on it and when it is tossed, the initial spin imparted. Similarly, it is possible to statistically analyze the behavior of the atoms in a gas, even though the motion of its atoms and the collisions between them are governed by classical mechanical laws.

Many Worlds

We discussed earlier how the issue of measurement occurs because a literal implementation of quantum physics results in the superposition condition of not only the photon but also the measuring apparatus, such that we have a cat that is both alive and dead in the case of Schrödinger's cat. It turns out that avoiding it is one way to avoid this problem. Suspending disbelief, let us see what happens if we take the above scenario seriously and ask how we could say the cat was in such a state.

The reason we know that a particle is in a superposition of being in one slit and being in the other, going through a two-slit apparatus, is that we can establish and observe an interference pattern. However, to do the same thing with the cat, we will have to put together the wave function representing all the electrons and atoms in both live and dead cats to form an incredibly complicated pattern of interference. This is a totally unrealistic assignment, in fact.

CHAPTER 8: CONCLUSIONS

The twentieth century may well be referred to as the quantum age. One hundred years after Einstein discovered that light consists of fixed energy quantities, how far have we come, and where will we go? This chapter aims to collect some of the earlier chapters' threads, to put them in historical context, and to make some guesses about what could be in store for the 21st century.

Early Years

For the first twenty years or so, after Einstein demonstrated the photoelectric effect in 1905, progress was very slow. However, once the wave-particle duality theory and its mathematical development were developed in the Schrödinger equation, they were easily applied to elucidate the atom's structure and its energy levels.

Within another twenty years, quantum mechanics had been successfully extended to a wide variety of physical phenomena, including the electrical properties of solids (Chapter 4) and the atomic nucleus's fundamental properties. In the late 1930s, the probability of nuclear fission (Chapter 3) was recognized, and this led to the first nuclear explosion in 1945, less than 20 years after his equation was first published by Schrödinger.

Since 1950

In the growth of our understanding of quantum physics concepts and applications, the discovery of quarks, which are now part of the modern particle physics model, was one instance of this. This resulted from the results of experiments involving very high-energy collisions between fundamental particles, such as electrons and protons; to answer the issue of the internal structure of the proton and neutron, it applied the principles of both quantum physics and relativity.

Much as an atom or a nucleus may be excited into higher energy states when fundamental particles collide at extremely high speeds with each other, similar excitations occur. It is possible to think of the results of these collisions as excited states of the original particles and the fields associated with them, but the changes in energy are so large that the related relativistic mass shift may be many times the mass of the original article.

As a consequence, excitations to such states are often thought of as producing new short-lived particles, which in a very short time, usually 10-12 s, regain their original form. The design of the machines needed to conduct such simple experiments required effort and cost approaching that of the space program.

During the second half of the twentieth century, development in the practical application of quantum physics was also tremendous. The invention of controlled nuclear fission (Chapter 3) quickly led to the creation of the nuclear power industry, which now produces much of the nation's electricity in some countries (over seventy-five percent in the case of France). A much greater problem has turned out to be the civil application of fusion, but research has now taken us to the point that this could soon be a real possibility.

In the last quarter of the twentieth century, the information revolution arising from the production of semiconductors and the computer chip (Chapter 5) took place and has undoubtedly been as dramatic and relevant as the industrial revolution two hundred years earlier. We can compute at enormous speeds, connect around the globe and beyond, and download information from the world wide web due to the quantum properties of silicon. Moreover, applying quantum physics directly to the processing of information (Chapter 7) has recently opened up the possibility of developing techniques in this field that are even faster and more efficient.

The Future

As far as fundamental physics is concerned, the more powerful machines currently being developed would allow the study of even higher-energy particle collisions: many expect this regime to break down the standard model of particle physics and to be replaced by another one that will create new and exciting insights at this level into the nature of the physical universe. Investigations into the behavior of matter at extremes of temperature and field will proceed in the area of condensed matter and could well yield new and fundamental manifestations of quantum physics.

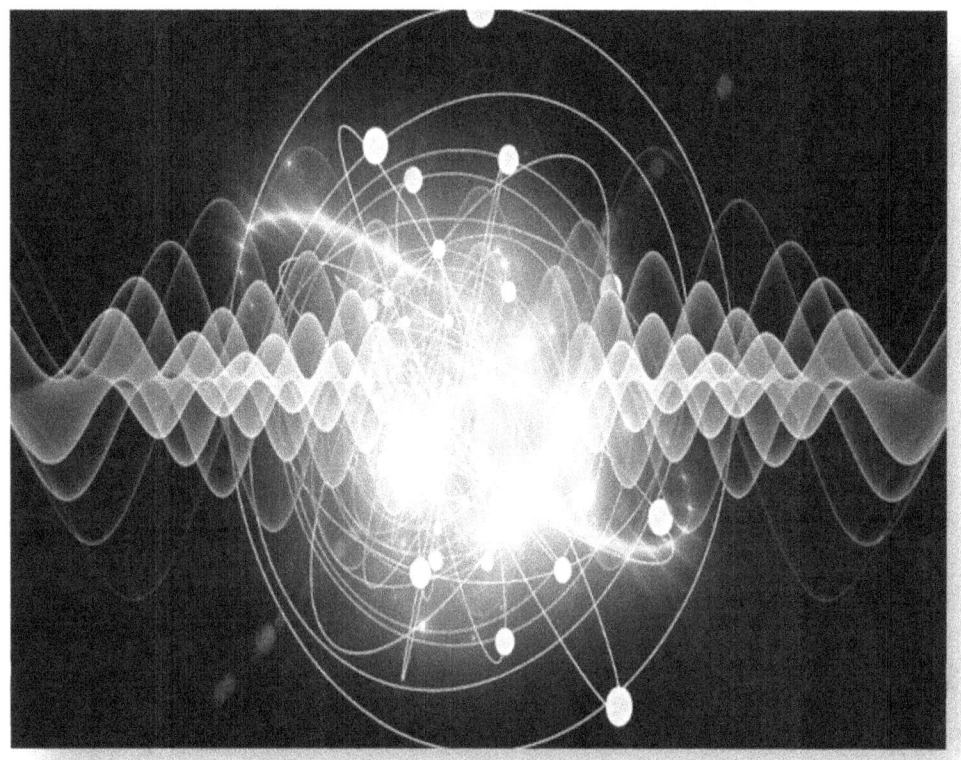

It would be perilous to forecast potential applications in quantum physics without a secure crystal ball. For several years to come, we can definitely expect traditional computers to continue to increase in power and speed: silicon's potential to surprise can never be under-rated. The superconductivity research will definitely proceed, but only very specialized applications seem possible unless and until malleable materials tend to remain superconducting up to room temperature. A major effort is currently being made to build devices for quantum computing (Chapter 7).

It is difficult to judge if this would work in the near future; it would be well advised for anyone thinking of betting on this happening to exercise considerable caution. Hopefully, very soon, the risks of continued fossil fuel burning will be better known, and the impetus to find alternatives will rise. The development of a new generation of nuclear reactors and advances in green technology, including those based on quantum physics, such as photovoltaic cells, could well result in this (Chapter 5).

The issue is so critical that we would do well to abandon debates about the advantages and drawbacks of the various alternatives: almost inevitably if we are to prevent a major disaster over the next fifty to one hundred years, all practicable methods will have to be exploited. The philosophical questions associated with quantum physics (Chapter 8) appear unlikely to be answered shortly.

Quantic physics seems to be a victim of its performance in this regard. The fact that such a large number of physical phenomena has been successfully explained and that it has not so far failed suggests that the debate is over alternate explanations rather than any need for new hypotheses. At least so far, any new way of looking at quantum phenomena that predicts outcomes other than those of normal quantum physics has proven to be incorrect.

In the future, a new theory might break this trend and, if it did, this would possibly be the most exciting fundamental breakthrough since quantum physics itself was invented. Perhaps such a development will arise from the study of the black holes' quantum properties and the big bang that our universe produced. New theories will almost certainly be required in this field, but it is by no means obvious that these will also answer fundamental questions such as the measurement issue. For a long time to come, the intellectual debate seems likely to continue.

I hope the reader who got this far enjoyed the ride. I hope you accept that quantum physics need not be rocket science and that you now understand why some of us have dedicated a large portion of our lives to trying to understand and appreciate what is undoubtedly the human race's greatest intellectual accomplishment.

www.ingramcontent.com/pod-product-compliance
Lightning Source LLC
Chambersburg PA
CBHW082215290526
45794CB00009B/3549